Laboratory Fume Hoods Explained
Second Printing

---

by
Chip Albright

A Fume Hood Certified Publication

Fume Hood Certified, LLC
PO Box 71477
Phoenix, AZ 85050 USA

www.fumehoodcertified.com

experts@fumehoodcertified.com

+1 623-662-0907

Disclaimer

This publication is designed to provide accurate and authoritative information with regard to the subject matter covered. It is sold with the understanding that the contents do not render professional advice. If you require professional assistance or expert advice, you should seek the services of a licensed professional.

Any company names, logos, or images used that are trademarked or copyrighted by others are acknowledged to be the property of the trademark or copyright owner.

experts@fumehoodcertified.com

Paperback: ISBN **978-1-7357110-8-9**

First paperback edition Octoberber 2020
Second Printing December 2025

Printed in the USA

Table of Contents

About the Author

Chip Albright

This book is a result of a 40 plus year journey into the world of laboratories. Over those years, Chip worked for a number of industry leaders in various roles. His introduction to ventilation products started in 1985 when he was transferred to the fume hood factory of Kewaunee Scientific in Lockhart, Texas. He designed his first fume hood in the late 1980s.

After Kewaunee, he joined Jamestown-JMP in Jamestown, NY. The first ASHRAE 110 was introduced in 1985, and a major revision was released in 1995. Much had changed as a result of advancements in fume hood testing, and VAV controls, giving room for innovation. Plus, with OSHA gaining jurisdiction over laboratories, many labs that had no fume hoods were now adding them. This was the perfect time to start with a blank sheet of paper and design the best hood we could.

That fume hood was known as the Isolator and included many innovative features including the chain and axle drive system for the sash and full pan steel superstructure. This product was the beginning of a long road of innovation.

In 1999-2000 Chip was Chairman of SEFA and a driving force for the rewrite of SEFA 1- Fume Hoods. Through this time, it became apparent that there was still ample room for advancing the art of fume hoods. Chip had moved to Collegedale, LLC in TN and took on the challenge to create the next generation of fume hoods. During this time the G^3 was developed. It was a high-performance fume hood with many advanced features. As these hoods were installed in large quantities, it became apparent that a great fume hood would only work as well as the laboratory ventilation system it was connected to. Chip developed a more holistic approach. The next few generations of the G3 focused on being more robust and capable of safe performance even on a marginal laboratory ventilation system. By 2010 it was becoming clear that there was a huge disconnect between those designing and producing fume hoods and the people who were engineering and installing laboratory ventilation systems.

Then there was that "aha" moment where he realized that fume hoods were not a standalone product even though that is how they were being sold. Instead they were the user interface to the mechanical ventilation system and until fume hoods became an integral part of the laboratory ventilation system, their performance was going to be questionable.

In 2008 Creative Solutions was created to explore how to best advance fume hood technology. In 2011, Chip became independent and began to explore the future of fume hoods. As a global consultant, he saw that in many parts of the world there was a total lack of understanding about how fume hoods worked and organizations were spending large amounts of money buying, installing, and operating fume hoods that were not providing an acceptable level of safety. After seeing the conditions that many users were working under, Chip became a user advocate and has spent much of the last decade educating users and laboratory managers about fume hood safety.

Lastly it became clear that because of the strong legacy of separation between fume hood manufacturers and engineers designing laboratory ventilation to achieve significant change, fume hoods were going to have to become smarter and less dependent on room conditions to perform safely. This has led to working on many innovative next generation smart fume hoods. Innovation has proven to be more accepted in countries like China and India where there is not as much resistance to change or regulation restricting innovation.

Chip has been a major player in creating standards though his work with SEFA and is a sought-after speaker on the subject of fume hood safety. His current focus is to continue to educate the market and the new generation of fume hood designers with the hope that meaningful change can take place in making laboratories truly safe for the user.

Introduction

Many laboratories have fume hoods installed. A chemical fume hood is a significant safety device that is critical to maintaining acceptable laboratory air quality and preventing user exposure to potentially harmful chemicals. Yet, the fume hood is the most misunderstood and misused safety device in the laboratory. So much mystery and myth exist regarding how a fume hood works and how to use it safely. It starts with the perception that a fume hood is a standalone device and if there is a fume hood in your laboratory and you use it, you will be safe. The reality is that a fume hood is only a small component in the mechanical ventilation system of a laboratory. It is the point of interface between the user and the laboratory ventilation system (LVS).

The fume hood does very little alone. Yes, there are good hood designs and not so good hood designs, but until connected to a properly designed and maintained laboratory ventilation system, the fume hood is just an enclosure that does nothing. The primary purpose of the fume hood is to protect users from exposure to harmful chemicals that are being used inside the hood. To do this, a fume hood must capture, contain, dilute, and exhaust those dangerous substances. None of this is possible without a ventilation system. Secondarily, the fume hood should provide some physical protection in the case of fire or explosion.

This book was written for informational purposes, it provides a background and explanation of how a fume hood works and how to use it in an effective way. The whole subject of fume hoods is full of misinformation, confusion and even controversy. Do a search for "fume hood books" and the only book you are likely to find was written by G. Thomas Saunders back in 1993. Why is there so little written about a subject as important as fume hoods? While there are standards and best practices, web posts and media articles that reference fume hoods, there is nothing from the 21st century that offers a comprehensive, factual discussion of the subject.

Having been in the industry for over 40 years I have seen it all -- the good, the bad, and the ugly. I have designed seven major fume hood product lines. I have worked for companies that have collectively built nearly 500,000 fume hoods over those years. Some of my later fume hood designs were truly world class and had great performance in the test laboratory. But it is possible to take that fume hood out of the test laboratory, install it in a working laboratory, and the same fume hood that performed so brilliantly in the test laboratory may fail to adequately protect users in a working environment. Users see the brand name and the test laboratory reports and understandably believe they are safe while using the fume hood.

In the post sales visits to laboratories, I often saw my world class fume hood not working properly. I spent a lot of time educating all stakeholders about fume hoods and how they must be integrated with the whole building to work properly. What I found was that due to the complexity of fume hood operation and the lack of understanding about how a fume hood works, no one really wanted to tackle the problem.

Through this book I hope to provide a needed source of accurate and timely information about fume hoods and their use in the laboratory environment. My goal is to cover the following points:

- the purpose of a fume hood is to capture, contain, dilute and exhaust harmful chemicals;
- fume hoods are simply a component of a laboratory ventilation system and do nothing alone;
- laboratory ventilation systems are about airflow;
- adequate testing – as manufactured (AM), as installed (AI), and as used (AU) is the only way to know if a fume hood is performing effectively;
- users are a major factor in fume hood performance and safety.

During the past ten years, I have worked mostly internationally outside the United States. Much of my time has been in India and China. It has been an eye-opening experience to see what is being done in these countries where there are few standards or regulations. Without government oversight such as we have in the U.S. with the Occupational Safety and Health Administration (OSHA), almost anything goes. One of the major problems in these countries is that many laboratories are designed and operated with no mechanical air supply systems. They use natural ventilation. In other words, they open the doors and windows when they operate the hood. It is impossible to manage air pressure and airflow balance without a mechanical supply air system. The other major issue is chemical storage. There are lots of chemicals in the open laboratory and it is common for many owners to shut the laboratory ventilation system off when users are not in the laboratory in order to save energy. The problem is that with no ventilation the chemicals in the fume hood and in the laboratory contaminate the laboratory air overnight and the users arrive back to a highly polluted laboratory the following day. The pollution is evident by the corrosion seen in these laboratories. I am quite certain that users have no idea of the hazards associated with these practices.

The last decade has been a real change for me. I went from being a supplier of fume hoods to being an independent consultant, trying to help companies make better products and to do more to protect users.

A while back, I was meeting with a laboratory owner who was preparing to build a new laboratory. The owner had been talking to several well-known suppliers about fume hoods. The plan was to use natural ventilation to save money. After some discussion, I said to him, "you do realize that these hoods will not perform safely without mechanical air supply?" His response was, "it doesn't matter if the hoods work, it only matters that we have them and that people assume they work." The indifference to user safety was shocking and convinced me to become a user advocate. It is the right thing to do.

Since 2010 I have given dozens of conference presentations and seminars on fume hood safety. I have worked with many companies helping them to improve their products and understand what it takes for a fume hood to perform effectively. I am compelled to share what I know with the hope that this information will help make a difference and that in the years to come laboratories around the world will be safer and healthier places to work.

Chapter 1

What Is A Chemical Fume Hood and Why Do We Use Them?

What is a chemical fume hood?

A fume hood is a safety device. It is classified as Personal Protective Equipment (PPE). Its primary purpose is to provide the user with protection against exposure to hazardous chemicals that they are working with and to provide some protection in the case of a fire or explosion.

Chemical Exposure is one of the invisible dangers of laboratory work. Exposure can be hazardous to your health. Although the symptoms of health damage from chemical exposure may not show up for years, the harm is occurring today. The Scientific Equipment and Furniture Association (SEFA) has published an excellent guide to exposure control devices such as fume hoods entitled, "Guide to Selection and Management of Exposure Control Devices in Laboratories." This guide is a good reference about chemical exposure and the devices and systems that can be used to prevent or minimize exposure.

Because a fume hood is a fire-resistant enclosure with a sash made from safety glass, it offers some physical protection from fire and explosions. Accidents are more common than you might think. Laboratory work can involve dangerous activities and therefore increased risk of physical harm. Training and preparedness for such events is essential to protect users.

Unless you plan on wearing a respirator while in the laboratory, you should be concerned about air quality. Breathing contaminated laboratory air can cause health problems, some of which may be permanent. Additionally, air quality not only impacts human health, it can damage equipment and compromise the outcomes of scientific work due to contamination of work samples. Maintaining acceptable indoor air quality is not only a best practice, but it makes economic sense as well.

What is a fume hood supposed to do?

The primary function of a fume hood is to prevent the laboratory users from exposure to airborne chemical hazards. To accomplish this, the fume hood uses air as its transport mechanism to:

- **Capture** the contaminants being generated or released within the hood;
- **Contain** them so they do not enter the lab;
- **Dilute** them with air to a safe concentration; and
- **Exhaust** them to the outside.

Are there different types of fume hoods?

Generally, fume hoods are designed to be used with a variety of chemical contaminants, but there are also a number of special purpose fume hoods. In a later chapter, we will look at each of these and why they are special.

- Perchloric Acid Hoods
- Radioisotope Fume Hoods
- Walk-In Fume Hoods
- Distillation Hoods

It is very important to recognize that not everything that looks like a fume hood is a fume hood. There is an industry standard definition of a fume hood. Products that don't fit that definition are often referred to as ventilation enclosures. In addition, there are other non-fume hood devices such as laminar flow hoods, chemical containment devices, ductless enclosures, ductless fume hoods, clean benches, gloveboxes, isolators, biological safety cabinets (BSC), and so on. In later chapters each of these will be discussed so that you can better understand why they are different. Using the correct product for your application is critical.

Where did my fume hood come from? Who selected my hood?

Choosing, purchasing, and installing a fume hood is a complex process. It can involve a number of people. Most of these people will not be involved in the use of the fume hood. It is uncommon for the people who will work with the hoods to have an active voice in the selection of product and even less common for them to be involved in selection or design of the overall laboratory ventilation system. Those often involved in the fume hood acquisition process include an architect, a laboratory planner, a mechanical engineer, a construction manager, the organization's chief executive or financial manager, a purchasing manager and maybe the laboratory manager. Generally, each of these parties have differing and sometimes even contradictory objectives. Because of the complexity of the overall laboratory ventilation system (LVS), it is rare that a holistic approach is used.

How was the fume hood selected?

Unfortunately, most fume hoods are selected as if they were stand-alone devices that work independently of other systems. They are selected more based on appearance than performance. Maybe the brand is important to the purchaser, but the reality is that the best fume hood in the world does nothing until it is connected to a laboratory ventilation system. A fume hood is not a stand-alone product, it is a component in the mechanical ventilation system. If a holistic approach is not taken when designing the entire system, there is a high probability that the fume hood will not perform as expected. In fact, what was purchased to be a safety device might actually become a safety hazard.

Who designed the laboratory ventilation system?

This question is actually more important than which fume hood was selected. Why? Because the fume hood will only deliver effective performance when connected to a properly designed and maintained mechanical system. Testing has shown that 50% of the hoods not performing at safe levels are failing due to issues with the Laboratory Ventilation system. Another 25% are poorly designed hoods, and the final 25% are lacking due to poor user work practices, but without a properly designed and maintained Laboratory Ventilation System, the best fume hood in the world will not work properly.

What is Supply Air?

Laboratory safety is all about airflow. To keep the laboratory/building in balance, the supply air must mimic the exhaust. Supply air is the air being supplied to the room from the HVAC system. It usually enters the room through an opening often in the ceiling called registers. Supply air is also considered creature comfort because it provides temperature and humidity control.

Most laboratories are under slight negative pressure, which means the air coming into the room, the supply air, is of a slightly less volume than the air being exhausted out. The lower pressure causes air to flow into the lab when doors are opened keeping contamination in the lab. For a fume hood to properly operate the exhaust and supply must track each other.

Understanding the requirement of air pressure balance makes it easy to see why fume hoods do not function properly with natural ventilation, such as open windows or doors. The laboratory doesn't have to be heated or cooled, but it does have to be mechanically pressurized for the fume hood to work properly.

Problems with air supply are the single biggest factor for a fume hood's poor performance -- whether it be the volume, pressure, or the method of delivery (cross drafts). Unless air balance is addressed and managed, the fume hood is likely not providing the anticipated level of safety for the user.

Many organizations have and are spending considerable amounts of money to purchase, install and operate fume hoods. Yet a sizable number of those hoods are not delivering the levels of safety they were designed for. Thus, the majority of this money is being wasted. Only when we look at the entire system in a holistic way will be able to address the issues that make many fume hoods unsafe.

Chapter 2

What Happens When A Fume Hood Isn't Working?

A fume hood is expensive to buy, install and operate. That expense is warranted only if it provides the intended protection for the user. It is easy to assume that just because there is a fume hood installed in a laboratory that it is providing a safe environment for the user. To determine if a fume hood is effective, you must understand what makes a fume hood work.

Loss of Containment

The primary function of the fume hood is to contain the chemical fumes within the fume chamber to minimize the risk that the operator will be exposed to contaminants from loss of containment. When there is loss of containment, there is a good chance the user will be exposed, either directly while working at the hood, or indirectly as the laboratory air becomes contaminated. Exposure to dangerous chemicals can have immediate effects, or it might take years for symptoms of exposure to appear. This is unlike biological exposure where the effect is more immediate. A smell of chemicals in the laboratory is a sign of possible loss of containment.

Lack of Dilution

Another failure is the lack of dilution. Lack of dilution has two components. Even hoods that are performing at an acceptable containment rate may not be effective at moving the chemicals into the exhaust stream. Many fume hoods have horizontal vortexes in the upper portion of the hood which can trap these chemicals and allow chemical concentrations to build up. This buildup can reach the Lower Explosion Limit (LEL) making the mixture explosive. If a spark is introduced to this mixture there will be an explosion or fire. In a well-designed hood, the vortex is single pass. That means that any chemicals picked up by the vortex will be exhausted after one revolution. This prevents the vortex from becoming concentrated with chemicals. There are also designs that minimize or eliminate the vortex.

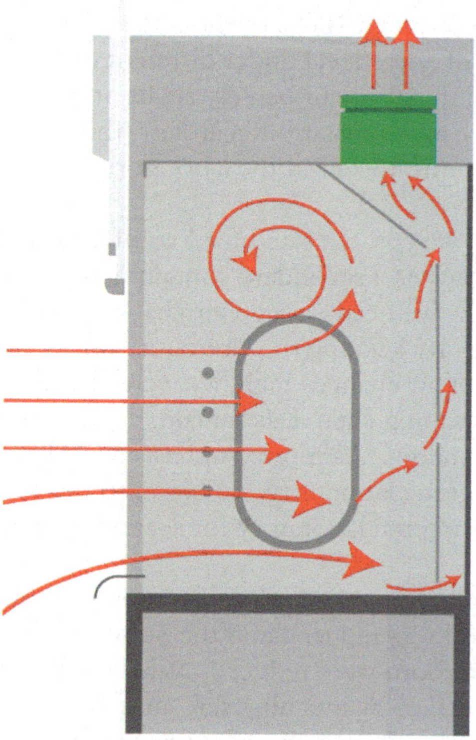

The second component is diluting the chemicals in the fume hood to a safe discharge level. The original intent of fume hoods was to capture, contain, dilute, and exhaust. In this scenario, the chemicals in the hood are mixed with air, usually room air, to dilute their concentration to a point that is safe to discharge. In the United States, the U. S. Environmental Protection Agency (EPA) sets regulations about the discharge of toxic substances. Generally, you want to dilute the chemicals in the exhaust to a safe level before discharging them. This is usually done by using air from the laboratory to dilute them. With some chemicals it is also possible to use a fume scrubber to clean the exhaust stream to a safe level. Discharging unsafe concentrations can be a hazard to maintenance personnel, to those downstream of the exhaust plume, and if there is reentrainment, to all the occupants of the building.

Laboratory Air Quality

Given the complexity of a laboratory building design, no fume hood will be able to contain 100% of the time. All hoods will suffer loss of containment at some level under different operating conditions. This loss of containment might be very slight or very infrequent or it might be very severe and constant. Regardless, the net result is the leakage of contaminants into the air in the laboratory.

In years past, we compensated for loss of containment by treating the laboratory space as a secondary containment chamber. For this to work properly, we need enough air changes to dilute the contaminants to a safe level. By keeping the laboratory under slight negative pressure and the rate of air changes high, we were able to ensure that the dilution of escaping chemicals remains at safe levels. But this strategy uses a lot of energy. As more emphasis was placed on the amount of energy used in the laboratory, we began to move away from the concept of using the laboratory for secondary containment.

To save energy, two things were commonly employed: face velocity of the fume hood was reduced and the number of air changes both in the hood and the room were reduced. While neither of these are bad strategies, it makes maintaining safe air quality a more complex and challenging task. Now to ensure safety the approach must be more risk based. What works in a low-risk situation might not be safe in a high-risk situation.

Chapter 4

How Design Impacts Fume Hood Performance.

Fume hood design is about the smallest of details. The difference between a good fume hood and a great fume hood is often only a few simple design features. Unfortunately, many fume hoods are designed by laboratory furniture companies rather than mechanical engineers. A major design objective by these furniture companies and lab designers is for the fume hood to coordinate with the appearance of the laboratory furniture. But a fume hood is a mechanical device not a piece of furniture!

Another problem is that many fume hoods are copies or even counterfeits of other products in the market. Even if the designer is copying a great fume hood, they usually don't have the technical knowledge to understand how the small details impact performance. As a result, they don't copy the small details correctly and what was once a highly effective fume hood becomes a mostly useless device.

To understand how fume hood design ended up in the purview of furniture companies you must look at the history of fume hoods. A more detailed history is provided in Chapter 15, but this is a brief version. The earliest fume hoods were adaptations of the fireplace and chimney. In Europe, the most common term is "fume cupboard" rather than fume hood. This is because the earliest furniture-like hoods were actually wall cupboards with glass doors placed over a window. When it was needed, the user simply opened the window. The glass doors kept most of the fumes from entering the room. Later, mechanical ventilation came along and fume hoods started to look much like they do today.

How did the modern fume hood design originate? To answer this, we first must look at the parameters that drove the designs. Height was dictated by the ceiling height of the laboratory. The hood height needed to be such that there was enough clearance for the sash to open and the ductwork to be attached. To get the fume hood into the lab, usually through a door, it was necessary to restrict the depth. The most common variable was the width. Generally, fume hoods were between

3 feet and 8 feet wide. These hood dimensions create a cubic space inside the hood. Most hoods are connected to the exhaust at the top and most work is performed in the lower half of the hood. From an airflow point of view, the space created by these dimensions is not optimal for containment.

Manufacturers often have a single model of fume hood that is available in widths of 3 feet, 4 feet, 5 feet, 6 feet and 8 feet. Even though they essentially look the same, each of these hood sizes have significantly different performance characteristics. Size matters. The fume chamber is a 3-dimensional space. The airflow within the fume hood chamber is also 3-dimentional. Changing the dimensions of the space changes the airflow. Altering the size of the chamber changes how the hood performs, yet most users would not know this. The model hood from the same manufacturer will have a different performance profile for each size. That is why it is important to have an (AM) As Manufactured test report for each size.

From the most basic point of view, containment is impacted by the airflow. Airflow patterns change with size, so if you change the size of the fume hood you change its performance. It is not a case of simply making the fume hood bigger or smaller, other details must be altered when the fume hood size is changed to minimize negative effects on performance.

How is airflow optimized? Observe how the air enters the hood, how it exits the hood, and how the airflow reacts at the various sash positions. The objective is to minimize and control turbulence and create an effective pathway for the air to capture the contaminants and then move them into the exhaust stream. The goal is to establish a smooth and mostly laminar airflow with minimum turbulence. Ideally, the fume hood design should eliminate the small eddies and the reverse flow and reduce or minimize vortexes both horizontal and vertical.

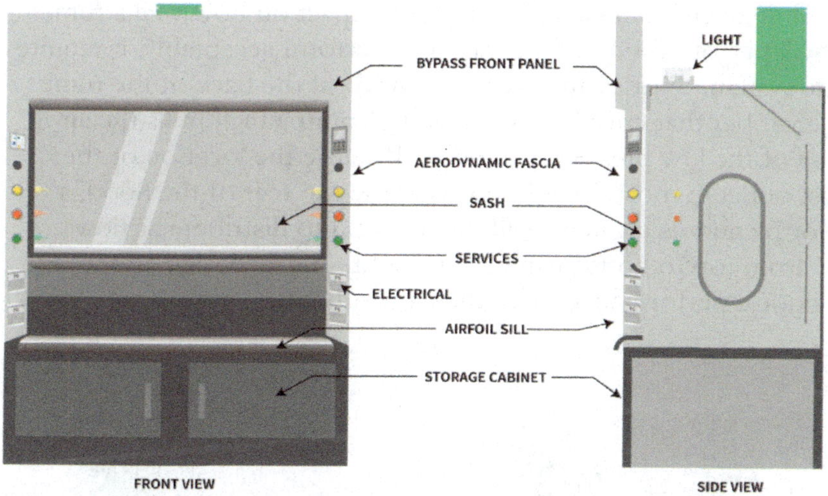

Observe how the air is entering the hood. All hoods should have a lower airfoil. The lower airfoil helps accelerate and curve the air at the bottom of the sash opening to sweep the worksurface and push contaminants to the back of the fume chamber to be exhausted. The upper airfoil has less function because it simply directs the air down the outside of the sash. When an upper airfoil is coupled with a well-designed sash handle, it directs a sheet of air inward towards the back of the fume hood. While these elements address the horizontal surfaces of the sash opening, the vertical components must also be considered. The vertical posts have an impact on how the air enters the hood. Curved surfaces are always preferred over flat ones.

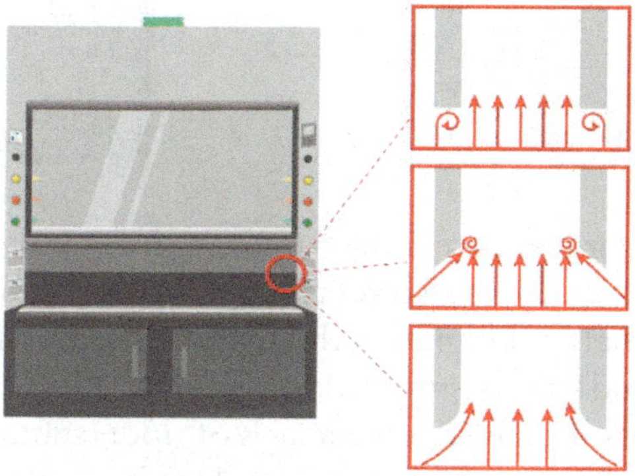

This shows the difference between a flat post, an angled post and a curved post. More attention is often to mounting the services than to the airflow.

Very subtle details can have a significant impact on how well a fume hood will perform. For a fume hood to perform acceptably, it requires a baffle system. Baffles are structures inside at the back of the fume hood chamber that create a pressurized plenum which pulls the air in because of the low pressure behind it. Because the location of the exhaust outlet in most fume hoods is at the top rear of the hood, a space or plenum is created to allow the hood to distribute airflow evenly from top to bottom and left to right. Baffle design affects a fume hood's performance more than any other design feature.

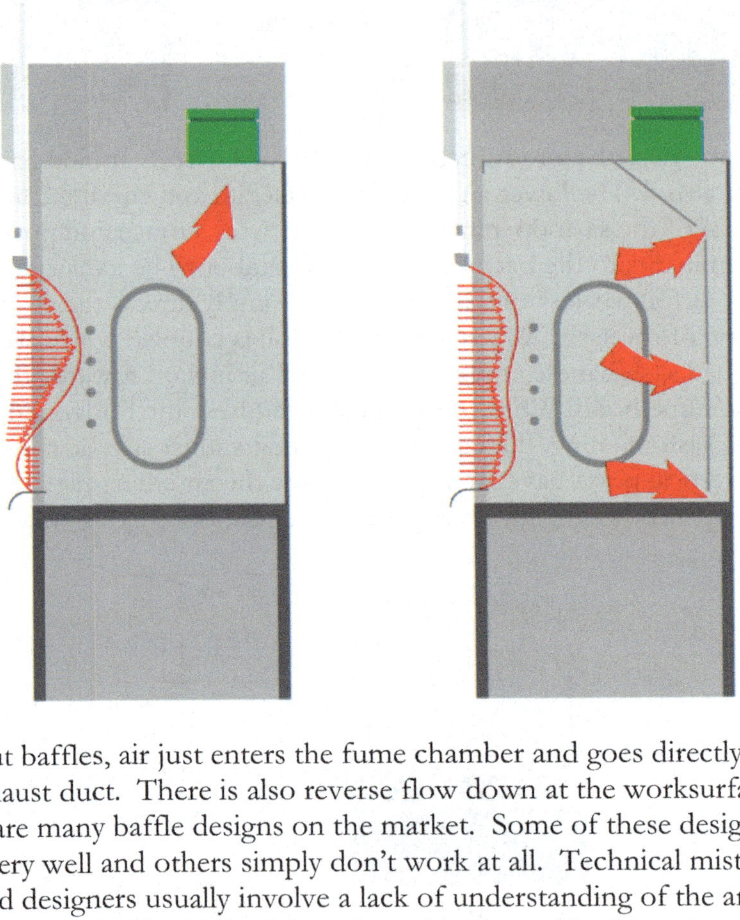

Without baffles, air just enters the fume chamber and goes directly to the exhaust duct. There is also reverse flow down at the worksurface. There are many baffle designs on the market. Some of these designs work very well and others simply don't work at all. Technical mistakes by hood designers usually involve a lack of understanding of the art and science of baffle design. Many fume hoods could achieve a dramatic improvement in performance by replacing the baffles. If a fume hood is not performing well and replacing it is not an option, upgrading the baffles might improve performance.

Because fume hood performance involves a number of design features, performance must be evaluated in a holistic way. While each feature may have a small impact, performance must be viewed in the totality. To develop a world class fume hood, the manufacturer needs the latest design tools such as CFD (Computational Fluid Dynamics) modeling and even more importantly, needs a world class test room with the latest instrumentation to validate the design.

As a buyer, it is essential to recognize that hood design is an art and a science and that some hoods are simply better than others. When considering a fume hood purchase, the more you know about how it works the better prepared you will be to make a good buying decision. While many fume hoods have good performance in a test room under ideal conditions, the same design may not perform as well under less-than-ideal conditions of the real world. What you want is a fume hood that is more robust -- a fume hood that will deliver acceptable results under a wide range of common operating conditions.

Outside of baffle design, the second most critical element is supply air. If a laboratory does not have mechanically supplied air, the fume hood cannot function properly. Natural ventilation is not appropriate for a laboratory with a fume hood. The variations in natural ventilation make it impossible for the fume hood to function properly. However, having mechanical ventilation for supply air does not automatically ensure that the fume hood will function properly. How that air is directed through the laboratory is critical. For example, the placement and type of outlets for supply air is important to minimize cross drafts that affect air currents.

The location of fume hoods when planning laboratory layout is often treated more like furniture than a mechanical system. Fume hoods are very dynamic and where they are placed makes a difference in their performance. Two of the most common placement sites for fume hoods are in corners and side by side. Such placement makes directing the airflow in the laboratory more difficult

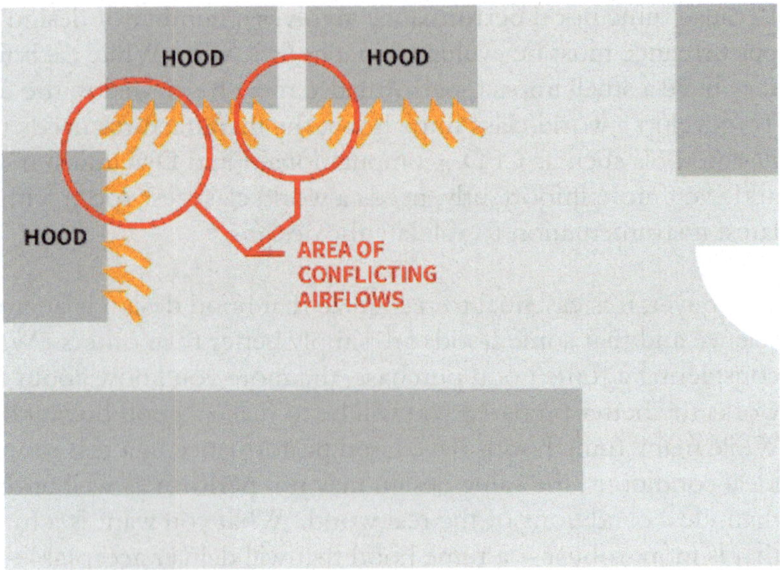

Because a fume hood's performance is so dependent on supply air and how the air is entering the fume hood, any air disruption can compromise a hood's performance. For example, locating a hood in an area that is a primary walkway or is near a door is not optimal.

Foot traffic and door swings contribute to disrupting room air patterns. The door should always open out from the laboratory for safety reasons. Should there be a fire in the hood, egress must be easily accessible. When the door opens, it allows a large volume of air to rush into the laboratory space because the air pressure in the laboratory is negative to the air pressure in the hallway. This rushing air has an effective velocity of from 2 to 5 miles per hour (175 to 450 fpm). This can not only cause cross drafts, but can also cause the room air pressure balance to change causing a shift in differential pressure and a corresponding loss of fume hood containment.

A person walking both pushes and pulls a significant volume of air. This movement disrupts the air entering the hood and can even draw air out of the hood. For this reason, fume hood locations should be away from areas of heavy foot traffic.

Hood diversity is a concept that sizes a ventilation system with less capacity than the sum of peak demand to reduce system size and energy usage. Systems are designed with a Diversity Factor, which is the theoretical maximum exhaust airflow quantity that is required at any point in time. For example, if a 50% diversity factor was used, it is assumed that no more than 50% of the fume hoods will be operating at any one time. You might ask why would we do this? The answer is simple, by incorporating a diversity factor, it enables downsizing HVAC system components and thus results in a smaller capacity ventilation system. The overall intention of applying a diversity factor when designing a VAV (Variable Air Volume) ventilation system is to achieve a lower life cycle cost. A smaller system costs less to operate.

Designing with diversity is an acceptable practice but it must be done with care. Designing for diversity differs from designing for flexibility. Systems designed for diversity may limit flexibility because they have only enough capacity to operate a predetermined percentage of laboratory chemical fume hoods at any one time. Both value engineering and sustainability can drive designs to 'right' size fans, ducting, air handlers and associated equipment.

If your laboratory was designed with a diversity factor, then it helps to know what the factor is. A hood might be working great at one moment, then start losing containment because the diversity factor has been exceeded and there is not enough air to maintain the proper airflow through the hood. When the diversity factor has been exceeded, then all operating hoods suffer, not just the last one that came online. In the case of diversity, it is necessary to manage the sash position and keep the sash closed when possible to help reduce the amount of air needed. Keeping the sash closed or opened the least possible amount reduces the amount of air that the fume hood uses. Less air also means less energy used.

One of the few movable parts of a hood is the sash. A good working sash is a safety feature. The sash acts as a safety shield. When working at the hood, the sash should always be closed below the face and the breathing zone. The more closed it is, the safer it is. In case of a runaway experience or a fire, the sash should be completely closed. Therefore, having a sash design that moves easily with one hand allows the sash to be closed quickly without having to stand directly in front of the hood.

There are several sash designs. Given how the air enters the fume hood, a vertical rising will provide the best performance. Use of any other design should require a very compelling need. There are a number of combination sash designs in use. These have a vertical rising frame to allow the hood to be fully open for setup. It also has horizontal panels that can be opened left to right for operation. While many people claim this sash style is safer, it isn't. When the horizontal sliding panels are opened, they create vertical vortices. These vortices grab contaminated air from the center and back of the fume hood and bring it up the sash opening making it easier to escape into the laboratory or the user's breathing zone. Sashes are discussed in more detail in Chapter 12

Performance is dependent on airflow and anything that disrupts the air patterns in the laboratory will impact performance. An empty fume hood will test differently than a loaded one. Blocking the baffles can totally disrupt the air exhaust pattern and allow copious quantities of hood-contaminated air to enter the laboratory. The setup within the fume hood will impact its performance. All equipment should be at least 25 cm behind the sash. Large objects or equipment should be elevated 50mm above the worksurface. And the baffles should not be blocked. These guidelines must be emphasized during training. But because there is no easy way to monitor compliance, the ultimate performance of the hood will depend on good user discipline.

The takeaway is that fume hoods are complex mechanical systems and a holistic approach to their design, installation and use is needed to obtain the highest performance. Unless all the components are considered, the intended safety provided by the fume hood will not be obtained. Fume hood design and location choice must be combined with the design of the laboratory ventilation system, and the physical laboratory layout in a holistic approach to obtain the highest and safest performance.

Chapter 5

What Makes A Fume Hood Work?

Fume hood performance is dependent upon the controlled flow of air in a laboratory. It is air that makes a fume hood work. Without the proper amount of air a fume hood cannot function safely. But it doesn't stop with the volume of air, it also involves the direction of the airflow and the amount of turbulence. Remember that air and water follow the same laws of physics. Air behaves like a fluid. The biggest challenge with fume hoods is that they have three-dimensional airflow. Most visualization tools, such as Computational Fluid Dynamics (CFD) modeling, only show us two dimensional slices.

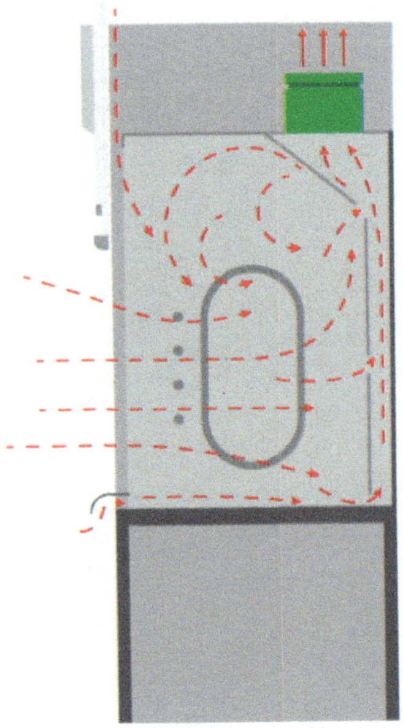

This graphic illustrates airflow. The lines represent airflow direction. If you take it at face value, then you would assume that the airflow at the bottom of the hood is laminar. But it isn't. Most CFD modeling lacks the computing power to fully model the airflow, so to get a reasonable simulation, the models decrease the precision and cancels much of the minor turbulence and eddies. This tool offers a way to validate some of the features, and tells us where to look for potential problems. But CFD modeling can only tell us so much about what the air is doing. Other testing such as smoke visualization and tracer gas containment testing are necessary to get a more complete picture of what is actually happening within the fume chamber. And the room conditions also impact the airflow within the fume hood. Much like a fireplace, there are many factors that determine where the smoke goes.

The challenges don't stop with an empty hood being tested. Most of the test protocols call for the fume hood to be tested empty. A loaded fume hood tests very differently from an empty fume hood. This can clearly be seen when doing the NIH (National Institute of Health) testing protocol which calls for boxes to be placed in fume hoods. But the testing protocol does not specify location. You can move the boxes around and see different results. Then when a person or mannequin is placed in front of a fume hood it changes everything yet again. To add to the inaccuracies, these tests are usually carried out in a near perfect test laboratory -- not in a working laboratory with less than ideal conditions.

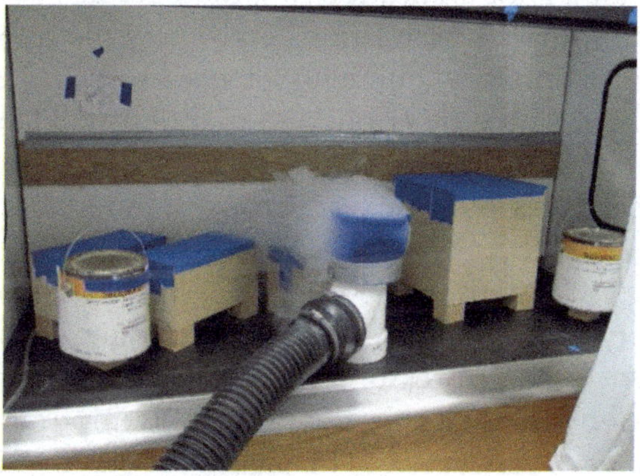

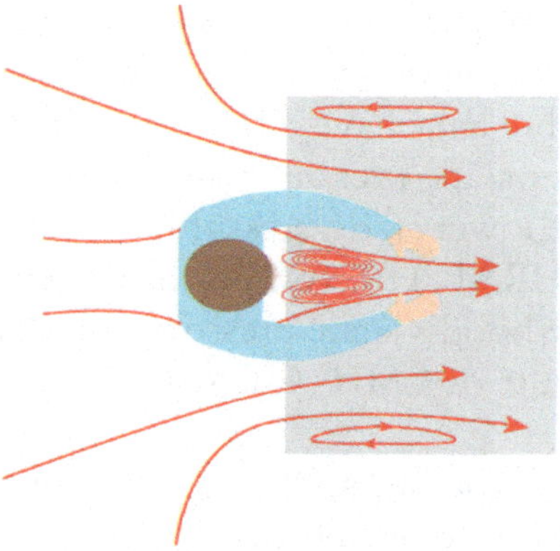

The person is much like an airplane wing and a low-pressure area is created in front of them generally causing reverse airflow.

If you are a fume hood user, have you ever worked with an empty hood or operated a hood without standing in front of it? Of course not. To be realistic, we need to look at the airflow under conditions that are more normal.

Once at a gathering of some the world's top fume hood experts, the comment was made," We don't design hoods to maximize containment, we design hoods to test well." There is a lot of truth to that statement. And if you know much about fume hood testing you know that the test protocols don't really address many of the real-world conditions. Good performance in a laboratory tested under perfect conditions, is not a good indication of a fume hood's overall effectiveness and there is no relationship to how the hood might perform in your laboratory. Fume hoods need to be tested or commissioned once they are connected to the overall laboratory ventilation system. If using the ASHRAE protocol, this would be the AI (As Installed) version of the test because a fume hood does nothing until it is connected to a ventilation system.

There are two types of pressures that are important in fume hood operations. One is static pressure. This is usually measured at the top plane of the hood. The reason static pressure is important is that it is the measurement of resistance. The higher the static pressure, the larger the exhaust system must be to create the desired face velocity. Face velocity is the speed at which the air is entering through the sash. A larger exhaust fan requires more energy to operate. So, if you have two fume hoods of the same size and you want to operate them at the same face velocity, but they have different static pressures, the fume hood with the higher static pressure will cost more to operate. The fume hood with the lower static pressure will be more cost effective. This is why the ASHRAE 110 -AM test has a very exacting method for measuring the fume hood's static pressure as well as total flow (CFM - Cubic Feet per Minute or CMH- Cubic Meters per Hour).

The more important type of pressure is differential pressure (DP). This is the pressure difference between two points. It also determines the direction of flow. Air flows from a higher-pressure area to a lower-pressure area. In the case of a fume hood, we want the airflow to be going into the hood, not coming out. We want to maintain a strong negative pressure within the fume hood compared to the room. Monitoring the differential pressure inside the hood chamber is a way to know if the hood is pulling air into the hood and maintaining containment. The stronger the negative differential pressure inside the fume chamber, the more robust the hood will be. A more robust hood will be less impacted by variating room conditions. A fume hood's design is a major factor in its ability to maintain adequate negative pressure required for containment. Hoods with less differential pressure between the hood chamber and the room are more likely to have containment issues. Most VAV (Variable Air Volume) systems add another level of complexity to maintaining an appropriate differential pressure. Today, ultralow flow face velocities are used as a technique to save energy. Ultralow face velocities also can make maintaining an appropriate differential pressure more challenging.

Recent studies have shown that air flowing over an object as small as 5-nanometers creates turbulence. A nanometer is one billionth of a meter which is much smaller than a human hair. The smallest details in a hood can cause turbulence. In most fume hoods you will find some laminar air and some turbulent air, from a performance point of view, the boundary layers between laminar and turbulent can be the most difficult to control, but controlling airflow to minimize turbulence is a critical element of good fume hood design.

If we could see the airflow it would be easy to determine if the fume hood was doing its job. We would be able to see all the air patterns and the disruptions. Seeing is believing. We could make adjustments until we had an optimal situation. Unfortunately, we can't see air, so we have to use other tools to try and understand what is happening.

There is a unique river management situation on the Ocoee River in east Tennessee (USA) that provides an illustration for understanding turbulence. In 1913 the river was dammed to provide hydroelectric power. After a lengthy legal battle, it was agreed that the middle section of the river (4.5 miles) was to be reopened for recreation daily and is home to some 20 Class III and Class IV rapids. For the 8-month rafting season, the original river is turned on and off daily. After sunset, the river is diverted to the powerhouse for use in generating electricity. At sunrise it is allowed to flow the natural riverbed so it can be used for recreation. I spent a number of enjoyable mornings watching the water rise to become rapids. When rafting the river at full flow it is hard to visualize what is under the water causing the uniqueness of each rapid, but if you watch the water rise, you get a much better understanding of how each underwater feature influences the flow of the river. Having designed many fume hoods, I often thought that if the air was as visible as the water, it would be easy to see what was really happening in the fume hood. If you could see the problems, they would be much easier to fix.

But given that air and most chemical fumes are not visible, determining what is actually happening inside a fume hood is a difficult task. While we have some visualization tools, in the end, it takes a lot of experience to really understand what you can't see. A fume hood looks simple, but it is anything but simple.

The fume hood is a dependent device. It only works as part of the overall laboratory ventilation system and is dominated by room conditions. We want the fume hood to perform under a wide range of conditions and situations, but a configuration that performs well in one situation might not do as well in a different situation. The complexities of airflow make the fume hood's job a difficult one. The more we know and understand about how a fume hood operates, the better we will be at maximizing performance and making the laboratory safer.

Users do not work in an empty fume hood. When you load the hood with equipment and stand in front of it everything changes. To evaluate safety, you must look at much more than face velocity. It is true that adequate face velocity is necessary for the fume hood to function safely, but there are many other factors that are more directly related to containment.

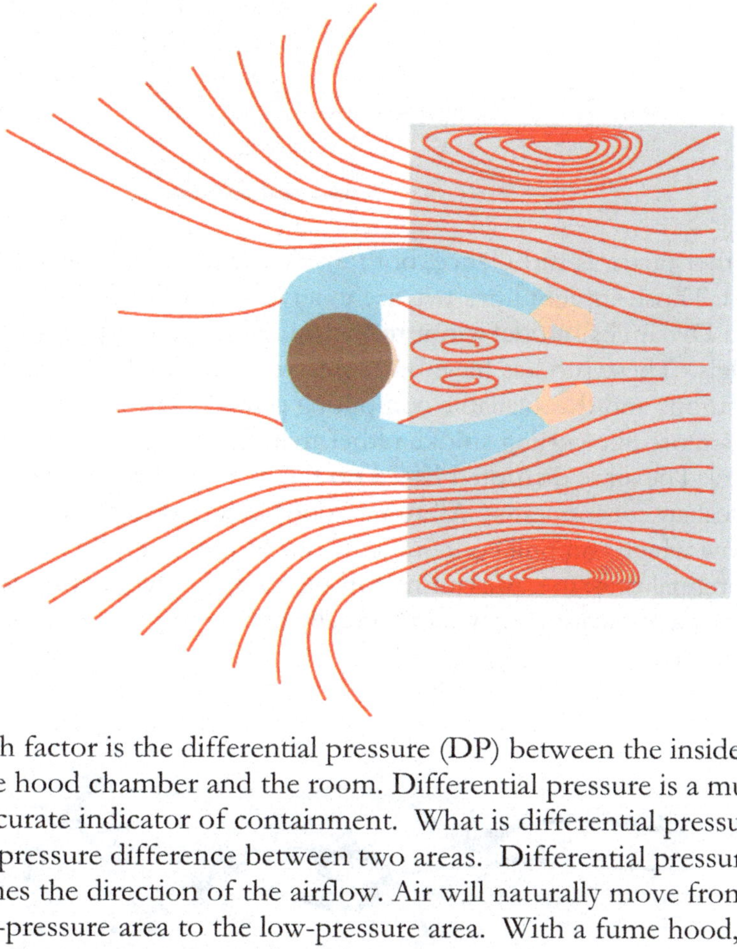

One such factor is the differential pressure (DP) between the inside of the fume hood chamber and the room. Differential pressure is a much more accurate indicator of containment. What is differential pressure? It is the pressure difference between two areas. Differential pressure determines the direction of the airflow. Air will naturally move from the high-pressure area to the low-pressure area. With a fume hood, we want the low-pressure area to be deep within the fume chamber to facilitate exhaust. But the room conditions cause the pressure to continuously shift. As the low pressure within the hood weakens, it is easier for the turbulent airflow to pull contaminants out of the hood. Monitoring the DP provides a better indication of containment than monitoring face velocity. Although DP alone is not the perfect indicator.

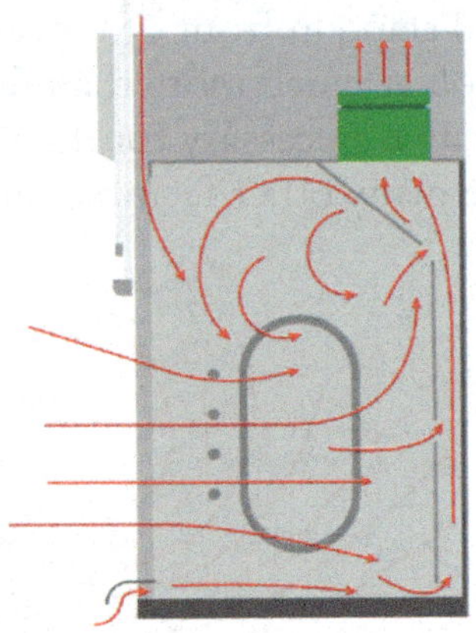

Another factor is turbulence, both inside the hood and outside the hood. Those straight lines you see in a typical CFD (Computational Fluid Dynamics) model is not really laminar as they appear in the models. These models have been greatly simplified based on model limitations and the available computing power. The airflow actually looks more like curling smoke rather than the straight lines in the model. The air is disrupted based on the mixing of the horizontal and vertical air components. All fume hoods have a degree of reverse airflow which can pull containments out of the hood during shifts of differential pressure. Measuring turbulence levels would provide an additional indication of what is actually happening to the air in the laboratory.

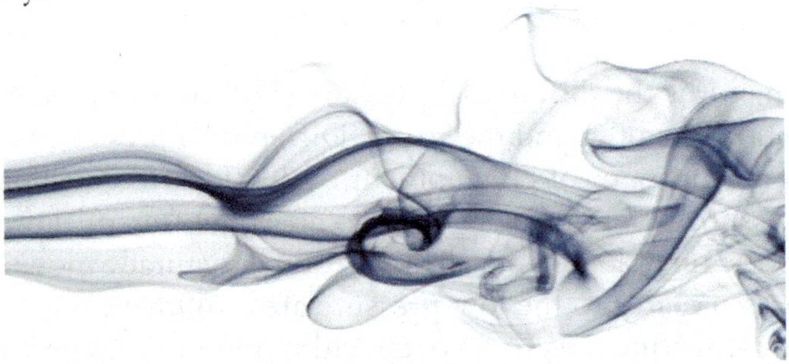

Fume hood tests such as ASHRAE 110-2016 AM and EN 14175-3-2019, measure fume hood effectiveness under perfect conditions. Conditions in real-world laboratories are not perfect. Additionally, in the test laboratory, only the properties of the fume hood are tested. Until the fume hood is connected to a laboratory ventilation system, there is no way to know how it will perform. Performing a test such as ASHRAE 110 – AI (As Installed) or AU (As Used) is the only way to know if a fume hood is actually achieving proper containment. Although there is technology under development that could change this.

Within the laboratory there can be a number of potential sources that can pollute the indoor air. The most likely source is a fume hood that is losing containment. Another is improper chemical storage. In addition, there is often work being done outside the fume hood on the countertops. Regardless of where the contamination comes from, the objective is to keep the laboratory air at a safe, breathable level.

This is done by replacing contaminated air with fresh air. This process is measured in terms of air changes. An air change occurs when 100% of the air in the laboratory is replaced with fresh air and is measured as ACH (air changes per hour). There are guidelines and standards that establish what an acceptable ACH is based on the type of space. These standards range from 2 to 20 ACH.

Typically, in laboratories, the recommended ACH range will be from 6 to 20. The exhaust rate and the size of the room will determine the ACH in any particular space. It is the job of the Safety Officer or Chemical Hygienist to determine what the ACH should be for a particular laboratory. If it has been determined that the desired ACH is 10, the cubic space of the laboratory is calculated and multiplied by 10 to determine how much air must be exhausted to achieve this number of air changes. Most laboratories have two components of exhaust, the fume hood and the general exhaust (GEX). To achieve the prescribed ACH the air must either go through the fume hood or the general exhaust.

Air being exhausted through the general exhaust adds little to overall safety. In prioritizing the method of exhaust, air exhausted through the fume hood(s) is preferred over general exhaust. When a fume hood operates at a lower face velocity, there may not be enough air exhausted through the fume hood to achieve the required ACH, making exhausting through the general exhaust necessary. Generally, fume hoods with lower face velocities are not as robust as the same hood with a higher face velocity. So, when choosing a fume hood, it is important to know what ACH you seek to achieve. Then, consider the size of the room, the number of fume hoods that will be operating in the room and their likely CFM. This should help you decide the best strategy. Remember, it is not the fume hood air usage alone that drives cost. It is the amount of air that must be exhausted to achieve the needed ACH that drives cost. A better fume hood that minimizes loss of containment may be a justification for lowering the ACH and accordingly, operational cost.

An energy saving strategy is to have a dual mode system. This type of system has a day mode with higher ACH and a lower ACH night mode. The practice of turning the laboratory ventilation system completely off at night and on weekends is not a good practice. An airflow of zero is not productive.

While it is very appropriate to lower the exhaust rate when the laboratory is not in use, turning the ventilation system off allows the quality of the air in the laboratory to deteriorate overnight. It may take hours after the ventilation system has been turned back on to get the contaminated air in the laboratory to a safe level. In laboratories (mostly in India) that have employed complete shut downs of the ventilation system when the laboratory is not in use, you can see signs of corrosion throughout the laboratory from the exposure of equipment to contaminated air. If the chemicals in the air are causing corrosion to painted steel and stainless steel, what are they doing to people's lungs?

Fume hoods should be the primary exhaust device in the laboratory. They are more productive in keeping the air safe than is general exhaust. ACH drives the amount of air that needs to be exhausted, but the fume hood should be the default exhaust method.

The final area impacted by face velocity is dilution. There are two factors of dilution to consider. The first is the type of contaminants that are being generated or released in the fume hood. The amount of air necessary to adequately dilute the contaminants in the fume chamber to safe levels will vary depending on what substance and how much of it is being created or used. But generally, less air and less face velocity results in less dilution.

There must be sufficient air entering the fume hood to dilute these chemicals to a safe level before discharging them outside into the environment. As the amount of air entering the fume hood is decreased, the dilution rate may be lowered to an unsafe discharge level.

The second factor of dilution to consider is the fuel to air ratio. When there isn't enough air to provide adequate dilution, it is possible for the mixture to become explosive. Even dust or powder can become explosive when the fuel-to-air mixture is right.

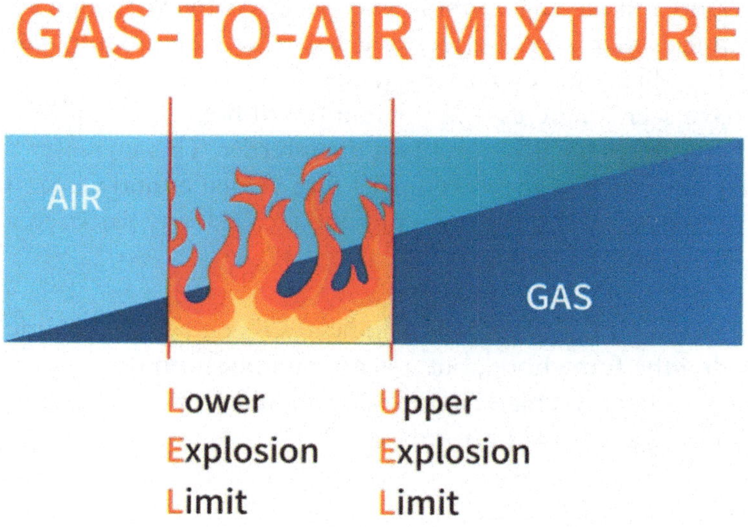

Conditions likely to cause a fire or explosion are described as the "lower explosive limit" or LEL. This is defined as the lowest concentration (by percentage) of a gas or vapor in air that is capable of producing a flash of fire in the presence of an ignition source. The LEL depends on air volume. The amount of air mixing with the chemicals determines the concentration by volume. The safest condition is for the mixture to be too lean (much more air than fuel) or below the LEL. If a lean mixture is exposed to an ignition source, nothing happens. But as the concentration builds, the mixture becomes explosive. Even as the concentration passes the LEL range and becomes too rich (more fuel but less air), it can still burn causing a fire. Less face velocity means that there is less air in the fume hood for dilution.

When looking at the LEL there is another issue that is impacted by fume hood design. Some hood designs with a strong upper vortex can develop higher fuel concentrations within the vortex. A fuel rich mixture in the vortex of a fume hood can easily cause an explosion. The design issue here is how effectively the fume hood evacuates the vortex. In some fume hood designs, the vortex just keeps turning round and round and picking up more chemicals. As the vortex picks up more chemicals that are not effectively exhausted, the mixture may cross the LEL and become explosive. While there are no formal studies, we have seen an increase in the number of hood fires and explosions in recent years as reported on the internet. This increase coincidentally correlates to the lowering of face velocities and CFM. It has therefore been surmised that lowering the face velocity has been a factor in this increase.

It bears repeating that the face velocity indicates that a certain volume of air is entering the fume hood, but it is not an indication of containment or safety. Average face velocity should only be used as one indication of fume hood performance.

Chapter 7

Loss of Containment

Loss of containment occurs when contaminants inside the fume chamber escape into the room. When a fume hood has a loss of containment, the fugitive chemicals escape into the laboratory and mix with the room air causing it to degrade. While there have been studies about the impacts of indoor air quality on human health, none have specifically focused on laboratories. Generally, these studies have focused on CO_2 levels and cleaning chemicals used in buildings. Laboratories add another level of complexity to the maintenance of indoor air quality because chemicals used in the laboratory can also become airborne.

If we use the classic definition of indoor air quality as a baseline, we can assume that the effects of fume hood loss of containment are additive to the other factors. Depending on what chemicals are in the fume hood and the lab, the impact of air quality on health varies. The second significant source of poor air quality is improper chemical storage.

Given that a fume hood is both a piece of personal protection equipment (PPE), and an exposure control device (ECD), it is clear that its primary function is to protect the user from exposure to potentially harmful substances that are inside the hood. Loss of containment is clearly a failure of the fume hood's primary function.

Also given that the fume hood is a major component in the overall mechanical ventilation system, the secondary function is managing and maintaining laboratory air quality. Historically, the fume hood was the primary containment device and the room itself was a secondary containment zone. Any loss of containment was diluted into the room air to keep the concentrations in the air at a safe level, but as ventilation technology advanced and a greater focus was placed on energy savings, this goal became more difficult to achieve.

In the 1990's, variable air volume systems were not common and the number of air changes in laboratories was quite high. It was generally accepted that the fume hood was going to lose containment from time to time and that sometimes work that should be done inside a fume hood was being done on the open bench. As a result, laboratory ventilation systems were purposely designed with a very high air change rate. These laboratory ventilation systems brought in lots of fresh air to flush the room. Although not as technically exact as modern laboratory ventilation systems, these systems probably did a very good job of providing clean air for the users to breathe, but with a high cost in energy used.

Air changes as measured in ACH are the real driver of laboratory energy costs. The more conditioned air that flows through the laboratory, the higher the energy costs. When considering safety and indoor air quality, the major factor is ACH. ACH is largely independent of the fume hood. The true energy costs of a laboratory ventilation system are associated with each air change – the cost to bring the outside air into the laboratory, to condition it, and then exhaust it. So how do fume hoods and loss of containment play into the overall costs of energy in a laboratory?

How can we justify fewer air changes and still maintain safe laboratory air quality? By looking at the risk of exposure we can evaluate an appropriate number of air changes. A better designed fume hood with less chance of loss of containment can be a justification for reducing the number of air changes. But somewhere in this search for ways to reduce the number of ACH needed, the balance between safety and energy got off center because of a failure to analyze the issues in a holistic way. As the emphasis began to shift to saving energy, we saw new concepts being put into place. Variable air volume (VAV) was one of the biggest. By controlling the amount of air a fume hood used, we could control how much additional supply air was needed to maintain air pressure balance between the fume hood and the room. To reduce fume hood air requirements, face velocities were slowed, which tends to increase loss of fume hood containment. In addition, the number of air changes was decreased, which means there is less fresh air to flush the laboratory. The air quality in many laboratories today is worse than what it was years ago.

intake for the supply air is also on the roof. Quite often there are also plumbing vents and vents for heaters or boilers. A lot of things are going on in the mechanical penthouse and on the roof. Again, this supports the need for a holistic approach to the entire mechanical ventilation system.

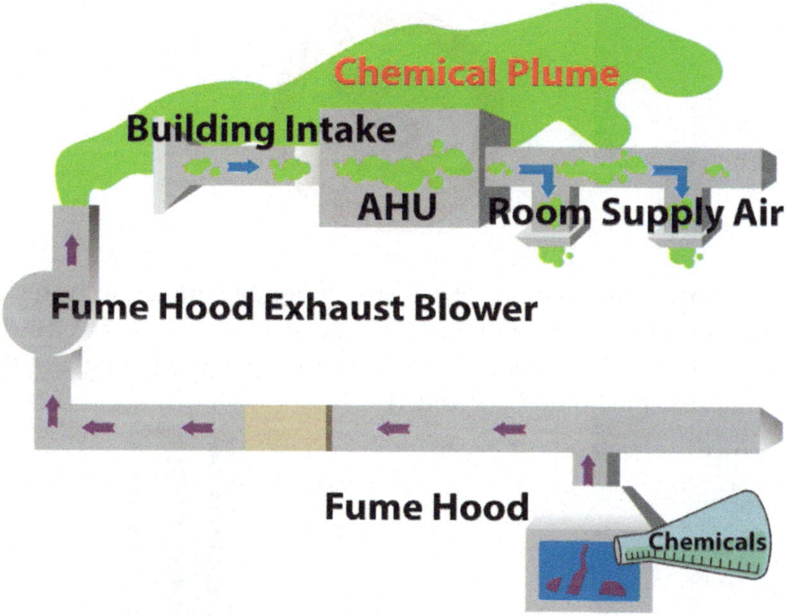

If the exhaust and supply air intake are not properly positioned and the stack height and exhaust velocity have not been optimized, the exhaust fumes can be sucked immediately back into the building through supply air intake as shown in the illustration above. This is the classic case for re-entrainment.

As a user, the most common method to detect re-entrainment is smell. When a smell of chemicals is detected in other parts of the building, you most likely have re-entrainment. Often re-entrainment is detected during routine fume hood testing. One test used to determine if re-entrainment is occurring involves releasing SF6 gas (sulfur hexafluoride) into a fume hood with the sash closed. Readings are then taken around the building, in the penthouse, on the roof, on the ground near the building and around any openings such as windows and exterior doors. This testing is very accurate and easy to perform.

Another common mistake is to place a traditional weather or raincap on the top of a fume hood stack. This defeats the purpose of using high velocity exhaust and tall stacks by redirecting the exhaust down onto the roof, almost guaranteeing re-entrainment.

To avoid this mistake, many stacks are designed with a rain water drain at the bottom of the stack where it connects to the blower. Others use a zero pressure weathercap. Whatever method is used, blocking the discharge stream should be avoided.

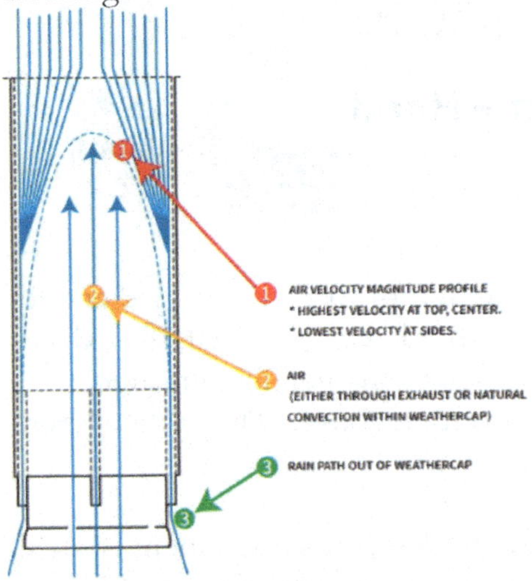

When you observe buildings and see discharge stacks on top, the fume hood stacks can be identified by their lack of weather caps. A lot of effort goes into keeping the indoor air chemical free, so don't overlook what is happening on the roof.

Here is where it becomes even more complex. Most fume hoods are VAV hoods and the volume of air going through the hood changes based on the sash position. If there are multiple fume hoods, the total volume depends on the sash position of each hood. If the exhaust demands are changing, then the air supply must change to keep the building and laboratory air volumes in balance. Therefore, the supply of air to each room must be variable and the DP (Differential Pressure) must be continually monitored to be able to maintain the room balance. All of this has an impact on the number of ACH necessary for the laboratory. If the minimum number of air changes are not being achieved by the exhaust devices, then air needs to be exhausted through the general exhaust. Yet it is preferable to exhaust through the fume hood(s) rather than the general exhaust.

Variable air volume requires constant monitoring to maintain balance. This monitoring requires sensors, valves and other complex components. It also requires ongoing calibration. Simply installing VAV equipment won't achieve the desired cost savings and safety standards. The VAV system requires ongoing monitoring, calibrations and maintenance. A well designed and maintained VAV system can help save large amounts of energy, but without the proper commissioning, testing, calibration and maintenance, it can be a huge waste of money and may not provide the level of safety anticipated.

From the standpoint of the fume hoods, a valid alternative to a VAV fume hood is a more simplistic approach. Simple constant air volume (CAV) hoods can be used. If good, high-performance fume hoods are chosen, they can be operated at a lower face velocity (60 to 80 FMP or 0.3 to 0.4 MPS). Equipping these hoods with an automatic sash controller and occupancy sensors allows the hood to adjust the airflow as needed. There is an "in use" mode and a setback mode for "not in use" or "nights." For the LVS, a two-mode system is used which has a high and a low setting. A day mode for when the laboratory is occupied, and a night mode for when the laboratory is not occupied. In night mode, the airflow rate might be reduced to 20 feet per minute (FPM) or a reduced number of air changes within the hood. While this is a form of air volume control, it is not as complex as a fully dynamic VAV system.

Before returning to the other LVS components and issues, let's refocus on DP (differential pressure). This metric is the best indication of flow direction. The goal is to have air coming into the laboratory from the supply outlets or when the door is opened. This air is being drawn into the hood when the sash is open, and then going out the exhaust. The differential pressure is changing constantly. Opening a door, opening or closing a sash, even walking in front of a hood, can create a localized pressure shift. Good containment in the fume hood requires the hood to maintain a constant negative pressure. These continual pressure changes can cause a hood to "breathe." It is possible for the hood to inhale and exhale which causes a temporary loss of containment. As the VAV system becomes more complex, there is often a lag time in the system sensing a pressure change and adjusting to it. It may take some time for the fume hood to become stable again. It is during this lag time that there is the highest probability of loss of containment.

Twenty years ago, "low flow fume hoods," as they were called at the time, were just getting the market's attention. A large U.S. project made the decision to use these hoods. The selling point was that these low flow fume hoods used much less air than standard fume hoods therefore would be saving energy and money. When the engineering calculations were done, they were right, the low flow fume hoods used much less air. While this seemed like a good thing, it created a huge problem for this particular project. There was a government requirement for a minimum number of air changes. Had the laboratory been equipped with standard fume hoods, the hoods would have provided enough exhaust with normal use. However, the low flow hoods did not. The project was forced to use the general exhaust to achieve the required number of air changes. While the fume hoods used less air, the laboratory didn't. So, the anticipated cost savings wasn't realized. It is important to focus on air changes first before making all the other decisions.

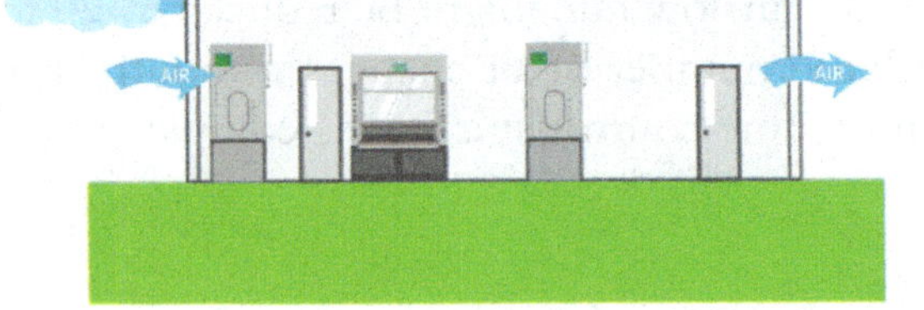

When we test for noise in a test laboratory, a sound meter is used to take a reading just below the exhaust duct without the hood connected, then the hood is connected and another reading is taken. The sound reading should not increase by more than 10db with the fume hood connected. Without the fume hood connected, most of the sound is created by high air velocity and turbulence within the duct. A noisy fume hood is actually caused by the ductwork rather than the fume hood. Air speed in the duct has a big impact on the noise level. The higher the duct velocity, the more noise, but an optimal duct velocity must be maintained. If it is too fast, it creates turbulence and noise and increases static pressure. If it is too slow, the contaminants in the airstream can fall out of the airflow. The optimal air speed in the ductwork is between 1300 FPM and 1600 FPM.

In some situations, dilution, through adding room air, may not be sufficient to adequately reduce chemical concentrations in the exhaust stream. There are two alternate concepts that can be used to further clean the exhaust stream. One is to use a scrubber, usually water based, which causes the chemicals to attach to the water, removing them from the discharged air. To a degree this just moves the pollution from air pollution to water pollution that may require waste water treatment. The other method is to filter the air with either HEPA or carbon-based filters. This actually removes the contaminants from the exhaust, but then the filters become hazardous waste and need to be incinerated. There is no perfect answer, other than to use less of the hazardous chemicals.

Now we are ready to dive deeper into some of the terms and concepts you might encounter. Refer to the LVS graphic on the following page (a repeat from the beginning of the chapter). Not every system is designed this way, but it will give you a general idea of how the components of a laboratory ventilation system work together.

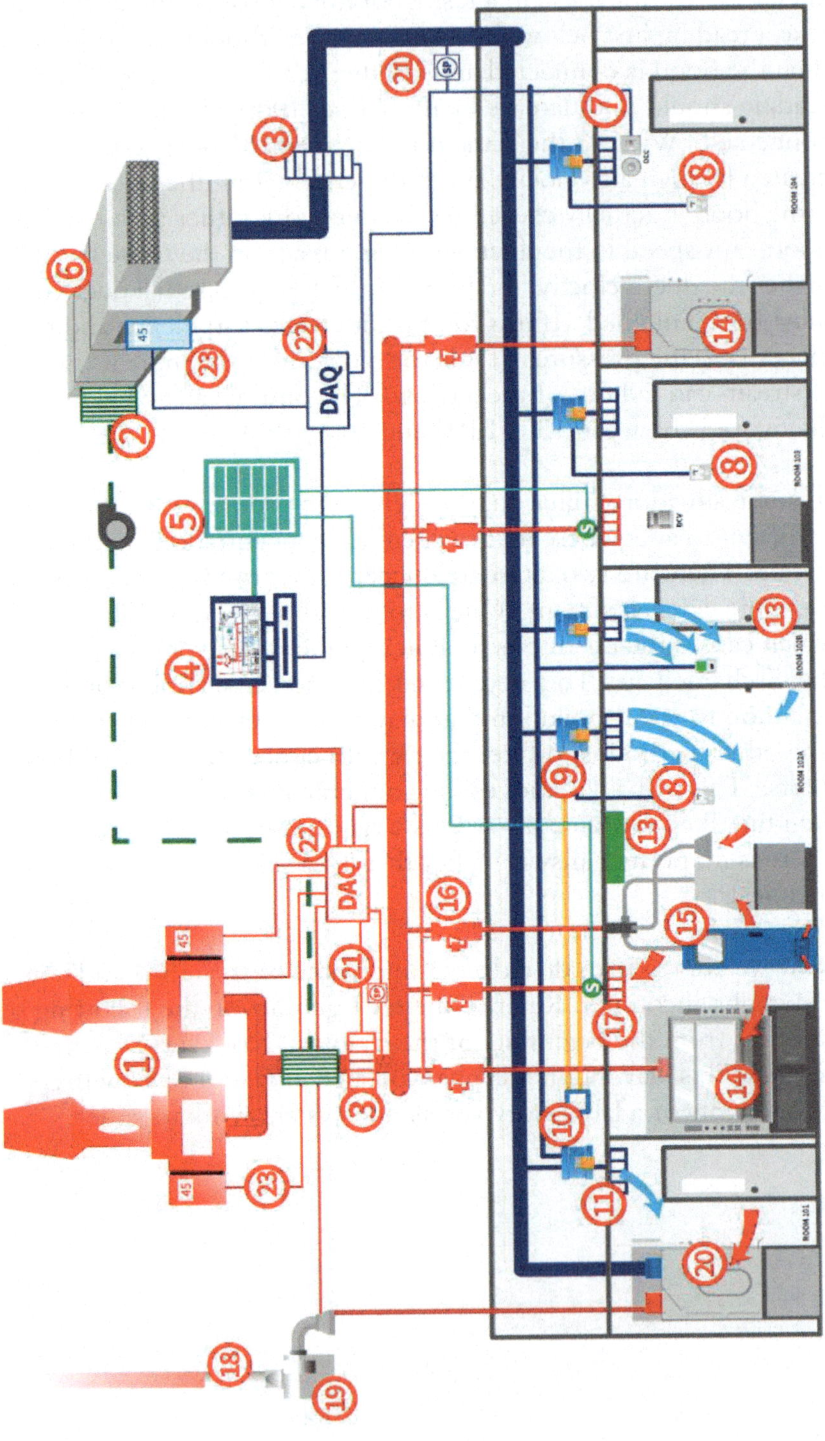
DAQ
DAQ
45
45
45
SP
SP
ROOM 101
ROOM 102A
ROOM 102B
ROOM 103
ROOM 104
BCV
DCC

Dual exhaust fans (1) – there are two things here to note, both the dual blowers and the type of blower. These are induced flow, venture type, exhaust fans and stacks with continuous power back up. Redundant venturi fans enable better plume discharge at lower flow and better system dependability. The combination of the variable frequency drive (VFD) and outside air bypass dampers (OABD) optimizes the ability to reduce flow, but will maintain sufficient plume discharge.

Energy Recovery System (2) – Various types of energy recovery systems can be used depending on the hazards in the exhaust.

Flow Monitors and System Sensors (3) – Exhaust and air supply systems often use flow stations and static pressure sensors. The measurement of total flow in addition to system static pressure improves system flow tracking.

Building Automation System (BAS) (4) – The LVS is actually part of the entire building ventilation system. The building automation system can monitor operation, detect operational problems and trend operation over time to provide useful operating metrics.

Contaminant Sensing Demand Ventilation Controller (5) – This is a concept that samples laboratory air and changes the ACH based on the quality of the air. This is the main controller and it interfaces with detectors in each laboratory.

Air handling Unit (AHU) (6) – Air handling units provide 100% outside air supplied to the laboratories that handle airborne hazards.

Occupancy Sensors (7) – The occupancy sensors detect the presence of people in the laboratory and enable flow to be reduced or setback when laboratories are vacant. This flow reduction lowers the air change rates. This lower rate of ACH potentially allows for the accumulation of concentrations of contaminants in the room which could exposure users upon re-entering the laboratory prior to purging.

Temperature Sensor (8) – The temperature sensors (i.e. thermostats) detect room temperature and translate a signal to the air supply discharge temperature controller for maintaining room temperature specifications.

Supply Air Valve (9) – The supply air for each room has a VAV valve. When the pressure sensors detect a pressure change, the room controller sends a signal that operates the valve to allow the correct volume of air to maintain room balance.

Room Controller (10) – The room controller monitors a number of factors within the room, but its primary function is to control the supply air to keep the room properly balanced.

Supply Diffuser (11) – The supply diffuser is the opening that allows the supply air into the room. In the laboratory it is preferable to avoid the high velocity diffusers because they can cause cross drafts in front of hoods. There are a number of types and materials available for laboratory applications.

Anteroom or Vestibule (12) – This is a room that serves as a secondary isolation to prevent migration or escape of airborne hazards to non-laboratory spaces.

Airborne Contaminant Filtration System (13) – These are ceiling mounted air cleaners. They contain carbon filters and they remove many chemical contaminants present in the laboratory air. The air is drawn into the unit and cleaned and then returned to the room. It has a net zero effect on DP or room air volume balance.

Variable Air Volume (VAV) Fume Hood (14) – This exposure control device is a VAV fume hood with a VAV valve and controller.

Local Exhaust Ventilation (15) – This group of exposure control devices includes vented chemical storage cabinets, gas cylinder cabinets, snorkels or spot extractors.

VAV Exhaust Valves (16) – These devices are comprised of sensors, actuators and flow dampers to modulate exhaust flow to satisfy the flow requirements of the VAV fume hood or other exhaust devices in the laboratory.

General Exhaust(GEX) (17) – This is a room exhaust used to help balance the room and to achieve the desired ACH. When the fume hoods are not exhausting enough room air, the GEX exhausts to maintain balance at the specified air changes.

Exhaust Stack (18) – This is the exhaust from the fume hood blower. It should be a minimum of 10 feet above the roof.

Fume Hood fan/blower (19) – This blower is connected to only one hood and is controlled by a switch on the hood.

Synchronized Supply Fume hood (20) – This is an exposure control device also known as a chemical containment device (CCD). It has both exhaust and supply air both with VAV valves and a controller to manage both and maintain the correct differential pressure within the hood.

Static Pressure Sensor (21) – This sensor is measuring the pressure loss in the ducting system. This is used to help determine airflow.

Data Logger (DAQ) (22) – This device is taking data from the various sensors and often analog and converting to a digital format that a computer can understand. This is often the interface between the sensors and the Building Automation System (BAS)

Variable Frequency Drive (VFD) (23) – This type of motor and controller allow the motor to operate at various speeds. When the demand is low the motor can slow down and as demand increases the speed of the motor which correlates to volume of air that is being moved.

Chapter 10

The User Is A Critical Part of The System

What happens or doesn't happen in a fume hood can make it unsafe. You would expect people that work in laboratories, because of the very nature of their education and training, to be highly disciplined when it comes to all aspects of their laboratory operations. However, that is often not the case. I have observed fume hood users doing things that make it nearly impossible for a fume hood to work properly. In most cases, these laboratory workers simply don't know any better; they have not been properly trained on how a fume hood works and how users should interface with it to maintain safe operation.

Recent news articles quote studies claiming that as many as 45% of laboratory users say they received no training on laboratory safety, including the proper use of a fume hood. Like all things safety related, the user is a critical part of the system and the outcome. For instance, safety glasses don't work if the user doesn't wear them.

Fume hoods are aerodynamic devices. Their performance is influenced by airflow and turbulence. What happens in and around the hood impacts containment. Therefore, good operating procedures greatly increase the probability of containment and reduce the risk of exposure to the user and other laboratory occupants.

If there are air quality problems within a laboratory, you may not be able to change the fume hood or upgrade the laboratory ventilation system, but you can train users to maximize the equipment they have by adopting good work practices.

When fume hoods are tested AU (As Used), many hoods fail to perform at safe levels. About 25% of those failures occur because of user practices. The corrective action is to train the users and help them understand how to do their work in a way that maximizes the fume hood's ability to contain hazardous materials.
Many users believe that a fume hood is protecting them simply because air is moving through it. They have no reason to question the performance. They do not realize that activity in and around the fume

hood impacts its performance. Fume hood users need training and knowledge to be an active participant in safe fume hood operation. After all, it is their health and welfare that is at stake.

As I say in my presentations at industry conferences, safety is, "no accident." Working safely with a fume hood requires a plan-- knowledge of what you are doing and what could go wrong. For example, if a runaway experiment causes a fire inside the hood, do users know what to do? Are there standard operating procedures (SOPs) for this event? Do you close the sash and get the closest fire extinguisher? Do you call 911 (emergency services)? Do you alert your supervisor? There are many possible actions you could take and many different outcomes. Some of the outcomes are not good. To ensure the results you want, there needs to be a plan and the user needs to know, understand, and follow the plan.

The actions of fume hood users affect everyone in the laboratory. Air quality affects everyone. If there is an accident, it puts everyone at risk. It isn't just about your health and safety, when fume hoods fail to contain and contaminate laboratory air, that puts everyone in the lab at risk of exposure. The poor air quality can damage equipment and compromise the results from work being performed.

I have long supported user training and certification before a user is allowed to use a fume hood. While some laboratories have adopted training and certification programs, it is not currently a requirement. As fume hood technology becomes more advanced, it will be necessary for users to receive extensive training before the hood will operate for them. There are fume hoods in use that will not operate unless the operator has the proper credentials in the form of a RFID ID or biometric identification, for now, simple training is a step in the right direction. If we are not planning to succeed, we are likely to fail. Lack of proper planning and poor procedures is the major cause of laboratory accidents.

The following questions are just a few that a user should consider to achieve a safe fume hood experience:

- Is the hood face velocity adequate?
- Are the back-baffle slots clear of obstruction?
- Is the work positioned at least 6 inches into the hood fume chamber?
- Is my hood housekeeping good?
- Is the bottom front air foil in place?
- Does my sash operate smoothly?
- Am I keeping the vertical sash closed to a point below my shoulders when I use the hood?
- Am I keeping my horizontal sash partially closed so as to obtain the smallest possible, but reasonable, opening?
- If my fume hood is indicating a low-flow alarm, do I notify my supervision?
- Can I rely on the noise I hear when standing at the hood face to tell me that the hood is working properly?
- Do I know what to do if I have a fire in the hood?
- Have my health and safety personnel performed a qualitative or quantitative evaluation of my hood, and if so, when and what were the results?
- Do I smell chemicals in the laboratory when I first come in?
- Do I see rust/corrosion around the laboratory or in the hood?
- Do I know the risks or potential issues with the experiments or work I am doing?

- Have I been trained to properly operate the fume hood and did I understand the training?

- Does the laboratory have safety procedures? Do I understand them?

- If there is an accident, do I know what to do?

Years ago, I started a "train the trainer" program and it has produced good results. I try to create a high level of fume hood expertise in the companies I work with. Training is only effective when the trainers know and can disseminate accurate information. I can't say enough about how important it is to educate and train users so they can be an active and functional part of the system. Once education and training occur, management's job is to ensure compliance. Where training has occurred, the biggest failure comes from the lack of user discipline.

As a user, you are a key component in the system that is there to protect you. You can make a difference in laboratory safety. It is your health and safety that are at risk.

Chapter 11

Fume Hood Testing

The only way to know if a fume hood is performing effectively is to test it. While there is debate about which test is best, any testing is better than no testing at all. This chapter contains an overview of some of the more common tests and explains why testing is so important.

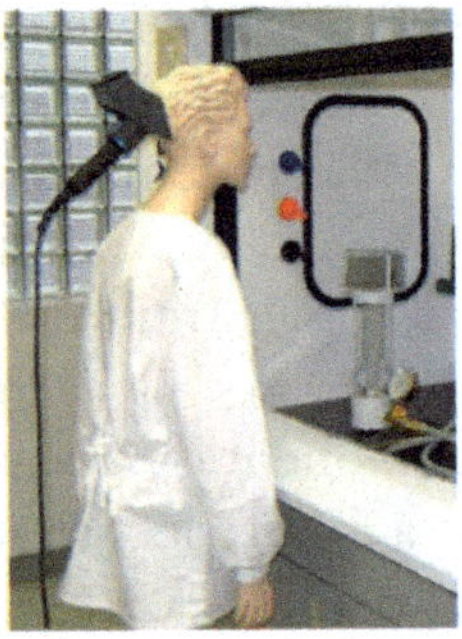

It bears repeating that a fume hood does nothing until it is connected to a laboratory ventilation system (LVS). The only way to know if a fume hood is working properly is to test it after it has been installed in your lab. A fume hood alone can't be tested because it is not a standalone device. It requires a mechanical ventilation system to function.

The laboratory mechanical design involves engineering not only the laboratory space, but also the entire building to work as a unit. That engineering makes a number of assumptions and contains supporting specifications. These specifications are used to select the various components that will be installed in the laboratory. The only way to know if these components are working together as specified is to perform the appropriate testing. This type of original testing is often referred to as commissioning.

There are few government regulations requiring fume hood testing. In biology and life science laboratories a device similar to a chemical fume hood known as a biological safety cabinet (BSC) is used to contain hazardous biological materials. The contaminant is biological rather than chemical. These BSCs are required to be inspected and tested annually, there is not a similar requirement for chemical fume hoods, even though a chemical fume hood is much more likely to have loss of containment than a BSC.

There are three different types of tests based on when and where the fume hood is tested. The first type of test is the AM or "as manufactured." This is generally a test that is done in a special testing laboratory. It is a test used to demonstrate that a fume hood could operate safely. The test laboratory appropriately creates a nearly perfect set of room conditions. This is because the AM inquiry is a test of the fume hood, not the room. To assess a fume hood's performance capabilities, you need to eliminate any impact caused by room conditions. Just because a fume hood can perform at a high level doesn't mean that it will. It is just one of many components in a very complex mechanical system.

The next type of test is the AI or "as installed." In this test, the fume hood's performance is observed once the hood is connected to the LVS. This is sometimes called commissioning and is the first indication that the fume hood is working as specified in the construction documents. This is not testing the fume hood independently, but rather testing how the hood performs when connected to the mechanical air system.

The final type of test is the AU or "as used." This test evaluates the performance level once the fume hood is loaded and a person is working at it. Again, you are testing the fume hood's performance along with the performance of the mechanical air system with the added factor of the fume hood being loaded as it would be in actual use with an operator in front of it.

These three types of tests can be performed in a number of ways to a number of different standards. Each type of test has a purpose and all three tests should be performed. Until the market provides new technology that gives the user a better indication of real-time performance, the only way to know if a fume hood is performing safely is to test it. Unless you test, you don't know if the hood is working or not.

When purchasing a fume hood, how can you tell if brand "X" is a better design than brand "Z"? It is very hard to know which product will perform best because the fume hood is not a standalone device. When a specialized test laboratory for fume hoods is built, the guidelines in the various standards are followed and the goal is to build the perfect room environment. This is done for two reasons. The first is to standardize test results. This standardization allows a purchaser to know each product was tested under the same test conditions so they can make equal comparisons between brands. The second reason for requiring the testing laboratory to be built to a consistent standard is to test only the fume hood. Testing all fume hoods in an equivalent environment makes the only variable affecting performance the fume hood itself.

Many fume hoods are designed without the benefit of having a fume hood test laboratory that meets these standards to test and optimize the design. If these fume hoods work well, its just because of luck, not science. If you want to design a world class fume hood, you need a world class test laboratory to test it in. That is why it is important for every company that is designing fume hoods to have a good testing laboratory. When purchasing a fume hood, you should ask about the supplier's test laboratory.

Why is designing a really good fume hood so difficult? The reason is that it is difficult to visualize and quantify what is happening within the hood. Years ago, there was a meeting of some of the world's best fume hood designers. The purpose of the meeting was to develop a new fume hood standard. As often is the case when you get a group of experts together, there is a lot of conversation. During this meeting, a very well-known expert said, "We don't design fume hoods for maximum performance, we design fume hoods that will test well." It is important to realize that each of these testing standards is evaluating something slightly different. Some fume hoods will test well regardless of how they are tested, but other hoods will test well on one standard and not so well on another. Why is this the case? Because the designers of those hoods were focused on optimizing the performance around a certain standard. Having tested many hoods using a number of different standards and tests, I would suggest that the more standards you test to (at the AM level) the better your chances of designing a highly effective fume hood. The more ways we challenge a fume hood in the test laboratory, the more robust it will be in the real world.

A well-designed fume hood that performs well in the test laboratory may not perform that well once installed in a real laboratory. The reason again is that the hood requires a mechanical air system to function. When AI testing begins, both the fume hood and the laboratory ventilation system together are being evaluated as a unit. A great hood will show a lesser performance when connected to a poorly designed LVS and a poorly designed fume hood might perform acceptably when connected to a great LVS. The ultimate performance will be a combination of the fume hood and LVS together. The logical first step is to acquire the best fume hood you can afford. The best fume hood is the one that has been tested to several of the standards and performs exceptionally well on them all. Starting with a fume hood that has demonstrated very robust AM performance provides the best chance of having an installed hood that performs well under a wide range of room conditions.

Once the LVS has been designed and the fume hood has been selected, AI testing will determine that the fume hood and the LVS perform together effectively and to the design specifications. Testing at this point holds the suppliers accountable for fume hood performance and failure to perform has to be corrected at the supplier's expense. Having a project commissioned is the only logical way to make sure the fume hood will perform as expected with a particular LVS. A great fume hood may not perform to specification because of problems with the LVS. Another reason to have an AM test for every hood model being installed, is that if the AI test results are poor, a comparison with the same AM test can reveal whether the problem lies with the fume hood or with the LVS. How can such a comparison reveal that? The reason is that the hood design didn't change, the poor performance is an indication that the LVS is the problem.

AI testing provides evidence that the entire mechanical system is functioning as designed and becomes a benchmark for future testing. AI testing is done when all mechanical systems are installed and the systems have been calibrated and balanced. This is done before the laboratory is occupied. Fume hoods are tested one at a time with no equipment in the hood. So, while this a very important test and should be done, AI testing is somewhat static and not reflective of the actual conditions in a working laboratory.

The next step in the life of a laboratory is actual operation. As the laboratory becomes operational, fume hoods are set up and used, people are moving around, doors open and close, sashes open and close. Building systems attempt to keep everything in balance. This is a much different environment from what was tested in the AI tests. If an AU test is performed and a fume hood tests poorly, unless AI testing was also performed, there is no baseline to determine the cause of the poor performance. Each test (AM, AI, and AU) builds on the previous test.

When a fume hood tests poorly in an AU test, you must ask, "Is the fume hood testing poorly because the LVS system has issues, or does it have more to do with the equipment setup or the work practices?" Once it is understood when and why to test, it's time to consider the various testing methods. There are many testing standards around the world. They range from very good indicators of fume hood performance to almost useless. Globally, two of the most commonly used standards are the ASHRAE 110 (current version at time of writing is 2016) and the EN 14175 (current version at time of writing was 2019). ASHRAE (American Society of Heating, Refrigerating and Air-Conditioning Engineers) has produced an American standard. ASHRAE is a worldwide professional association with members in 132 countries and it publishes many standards used worldwide. The EN (European Norm) is a European standard. It takes related standards from various European countries and normalizes them for use in the European Union. The EN standards draw from the old British Standard and the German Standard. For years there has been a debate about whether ASHRAE or EU standards are better. The real question isn't which is better, it is what is each of them testing and how are they testing? The test methods are very different. They use different methods to measure different functions. In the end they both measure some form of containment, but in very different ways. I suggest that fume hoods be tested by both standards because the challenges are different. Performing well by both standards is an indication that you have a robust product.

generated by a fog or smoke machine is another good way to visualize air patterns and movement. To further enhance this, pointing a handheld laser modified to have a flat beam into the smoke amplifies the visual observation of the turbulence.

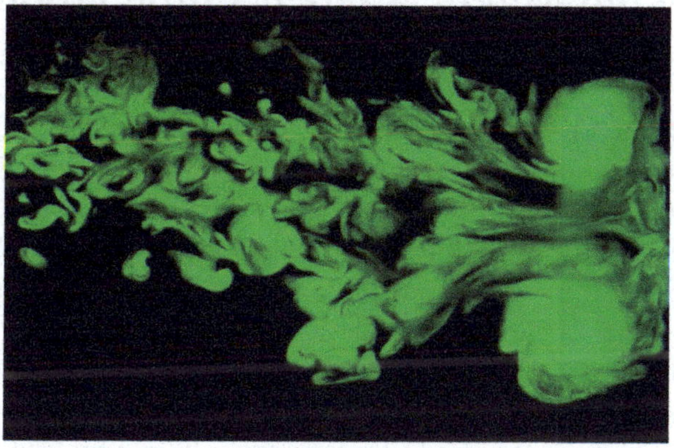

There is also a new technology soon to be on the market called Predictive Containment which uses sensors and artificial intelligence (machine learning) to monitor a fume hood's containment in real time. These "smart" hoods will be able to monitor the hood's containment continuously.

Another real-time way to test containment is to monitor the air quality just outside the fume hood. This concept acknowledges that the fume hood will likely have some loss of containment. When loss of containment occurs, this air monitoring system will detect fugitive chemicals and increase the number of ACH to dilute these contaminants to a safe level.

Lastly, it should be noted, that because SF6 has been identified as harmful to the atmosphere (greenhouse gas), there is a movement underway to change the tests that use SF6. The most talked about option is to change from SF6 gas to isopropyl alcohol (IPA).

The next few years will likely bring much change to the world of fume hood testing, including a more global requirement to perform regular testing, for now, any testing is better than no testing.

Chapter 12

Hood Design – What Makes A Great Fume Hood?

A fume hood's design makes a huge difference in performance. The more robust the design, the better the fume hood will perform, even under adverse conditions, but given the fact that the fume hood is dependent on the laboratory ventilation system to perform properly, why is design important? Laboratories are very complex structures making it difficult to manage airflow and balance. Laboratory room conditions create challenges for all hoods. The better the fume hood, the better it will perform under all these ever-changing conditions. I have tested thousands of fume hoods and observed even more. Some hoods are so poorly designed that you can tell, just by looking at the fume hood, that it will not work properly. Some designers simply copy features they have observed on other fume hoods without understanding the significance of the details of the features. These copies often don't perform as intended. Many fume hoods are designed by furniture companies that have designed the fume hood as if it were a piece of furniture, when it is actually a very complex mechanical device. Because air and water follow the same physical laws (fluid dynamics), a fume hood is more like a pump than a piece of furniture. If you designed a pump to move a liter of liquid per minute, to increase the volume to 10 liters per minute, you don't simply make the pump bigger. Yet with fume hoods that is often the approach. Fume hoods have a three-dimensional airflow and the airflow isn't linear; as the dimensions change so do the airflow patterns. Even the same model of fume hood in the different sizes will perform differently.

Why is it that fume hoods are most often designed by furniture companies when fume hoods are a significant part of the mechanical laboratory system? Wouldn't it make more sense if fume hoods were designed by the people designing the other mechanical components? The answer is deeply rooted in history. Most people trace the first modern fume hood back to the University of Leeds (UK) in 1923. It can be described as taking a wall cupboard and mounting it over a

turbulent, even at reduced face velocities. The reason some fume hoods perform better than others is solely dependent on their internal design. Conventional narrow depth "Walk-in" floor mounted fume hoods perform poorly because their Re numbers are double that of bench hoods, indicating twice the turbulence.

An error made by fume hood designers and industrial hygienist was to assume that because the face velocity was fairly uniform across the sash chamber opening, then so too were the airflow patterns inside the fume hood.

The airflow entering an enclosure will have a laminar profile but quickly become turbulent. This is exactly how a fume hood operates.

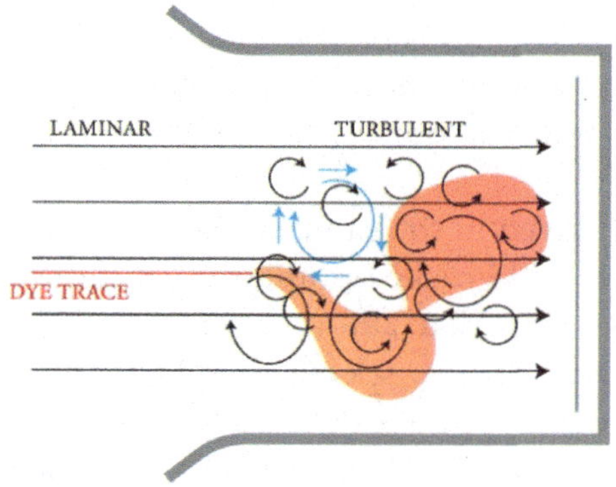

Users, as a critical part of the system, should have enough knowledge about the fume hood they are using to make informed decisions about fume hood safety. Their actions will impact the fume hood's ability to perform properly.

Size matters. The airflow in a fume hood is three dimensional. Because the Re number is affected by dimension, a 48" hood, a 60" hood, a 72" hood, and a 96" hood of the same model, from the same manufacturer, will have different performance results. This is because the airflow patterns are not the same. Every change to the hood has an impact on performance. Some changes make the fume hood better while others reduce performance. It is the minor details that separate a good fume hood from a great fume hood.

Where do the typical dimensions of a fume hood come from? Because fume hoods are most often designed and built by furniture manufacturers, the dimensions of fume hoods have been unduly influenced by the size of doors and ceiling height and the design of the laboratory spaces. The fact that a fume hood is aesthetically pleasing and fits through the door and onto a laboratory counter doesn't mean it will perform well. The dimensions of the standard hood don't necessarily optimize its performance.

Air is lazy, it will follow the path of least resistance. The good news is that airflow is manageable. The airflow inside the hood is three dimensional with a horizontal and a vertical vector component, and as these components collide, it produces air turbulence. Much like water, as the air flows across various physical features, the air is curved and bent and the speed and direction change. What might have started out as a smooth even flow can quickly become very turbulent.

If the fume hood stayed in the same state of operation and the setup never changed, we could easily optimize the airflow for great performance. Somewhat like automobiles or even bicycles, if they were only driven in the city at slow speeds then they could be optimized for that. As soon as the driver takes them off-road, everything changes. In the case of a fume hood, we have a sash that is always being opened and closed, the setup is changing and more importantly the room conditions are changing. Given the dynamic nature of fume hood operation, the product needs to be as robust as possible to provide the best chance for proper performance over this wide range of conditions.

After the size of the fume hood has been determined, what should a designer consider? First, how the air enters the hood. There are three surfaces to consider. The single most important surface is the lower airfoil. Next is the sash handle and then the vertical components which frame the sash opening. These are the left and right front posts, known by many names including columns, posts and facia. Each of these surfaces have an impact on how the air enters the fume hood.

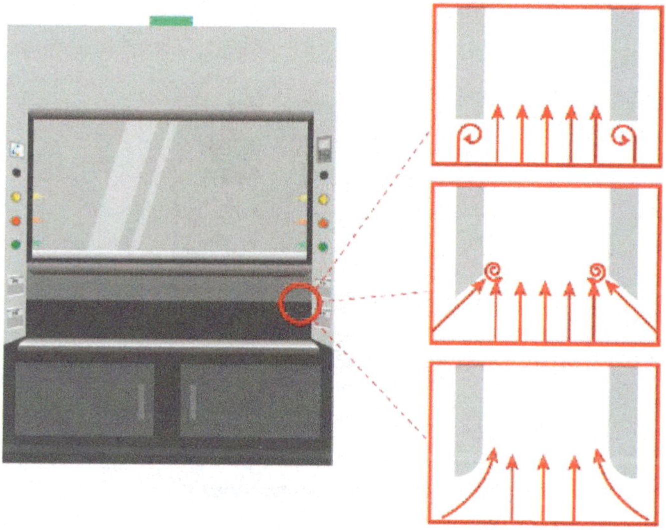

Next, what happens inside the fume hood? The air is pulled towards the back baffles because of the lower differential pressure, but this movement is not uniform. If you take a velocity probe and read the slot velocity across the back at the bottom of the hood, you will find it varies from left to right. You can do that at any other slot or horizontal opening and see many different readings. This uneven suction across the baffles causes the air to follow the easiest path. The exhaust duct is usually in the center of the top and the suction is stronger in the center. So more of the air entering the hood is pulled towards the center. Openings, or slots, near the top of the hood will have stronger suction than those near the bottom. Even if the airflow entering the hood was uniform, the further into the hood it goes, the more turbulent it becomes. The art of designing an efficient fume hood must take into consideration the placement and design of the baffles and their openings, and the pressurization of the plenum created by the baffle.

There are fume hoods on the market and hoods in development that have made improvements in balancing the flow going into the baffles. Some fume hoods actually have created a laminar flow where the air enters the sash opening and flows directly to the back of the hood remaining very laminar. Other designs have achieved a laminar flow at the top of the hood above the sash opening eliminating or greatly minimizing the traditional upper vortex.

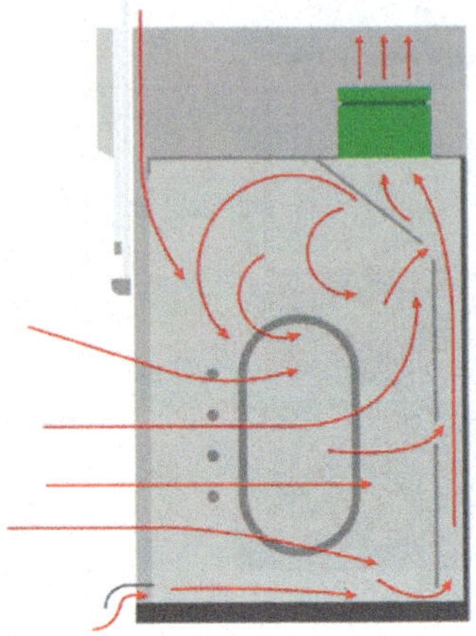

The more organized the airflow, the better the hood will contain contaminants, regardless of face velocity. The less turbulence in the fume hood, the better the containment. The best hoods are designed to be very aerodynamic with reduced turbulence and organized flow, but the only way to really know how well a hood performs, is to test it. As discussed in Chapter 11 on testing, some hoods test well using one testing standard, but not so well using another. The more standards or real-life situations you test against, the better you will understand the performance of your hood. Any visualization aids you can use will help you better understand what that invisible air is doing.

The majority of fume hood designs have a vortex at the top of the hood. This vortex is caused by air coming in the open sash and going to the upper back of the hood. This allows the air in the front top area of the fume chamber to just spin creating a vortex. As the sash opens and closes the vortex usually moves up and down with the sash. If the sash has been closed, the vortex has likely dropped nearly to the worktop. If the sash is opened quickly, it is possible for that buildup of contaminants to spill out before the hood can stabilize potentially exposing the user. Designs that either eliminate the vortex or prevent it from dropping down when the sash is closed will perform better and have a higher containment ability.

One of the last points worthy of discussion regarding fume design is the sash style. A single vertical rising sash will always perform better,

categorized into groups or classes. Here are a few of the hood types used in chemical laboratories:

Ductless Hoods

Ductless hoods are more similar to a BSC than a chemical fume hood but are intended for limited chemical use. Ductless hoods are filtered and they also recirculate the air. Depending on what chemicals are being used, these ductless hoods are outfitted with different types of filters.

Generally, these devices are intended for specific applications and not general use. If you are using a ductless hood, check with the manufacturer to verify that your application is appropriate for the device. Most manufacturers of these devices do an audit of the work planned for the laboratory before recommending use of a ductless hood.

Laminar Flow/Clean Bench (Horizontal Laminar Flow and Vertical Laminar Flow)

Laminar Flow/Clean Benches are generally used to protect the product inside the hood and not the operator. They are much more common in a cleanroom than in a chemical laboratory. These devices support very specific types of work. Often it is hard to tell exactly what these units are. They may look like a chemical fume hood, they may look like a BSC, or they may look like both. It is very important to know what type of device you are working with, what types of work the device was designed to perform and whether it is working properly.

Clean Air Hoods

Clean air hoods are strictly for product protection. They are not intended to protect the user in any way. Generally, they take room air, bring it through a HEPA filter, and blow it on the product. This creates positive pressure inside the hood rather than the normal negative pressure used for operator protection.

Chemical Fume Hoods

This is the class of products that this book focuses on. These devices are generally found in chemical laboratories or in places where chemicals are being handled and where the user could be exposed. A survey of the world market will reveal many brands of chemical fume hoods. In fact, it is not uncommon for a single manufacturer to have multiple brands or families of products. While the intent of all chemical fume hoods is the same, to protect the user from exposure, the approach used to accomplish this varies considerably. Also, the designs can be very different. Again, unlike the BSC and the Ductless Fume Hood, the Chemical Fume Hood requires connection to a laboratory ventilation system to work. The fume hood is merely the user's interface to the mechanical air system.

Traditional Bypass

The traditional bypass is one of the oldest types of fume hoods. It is a constant air volume hood. As with any hood, if the air volume being exhausted is constant, and the sash opening is reduced, the face velocity will increase. Without another way for air to enter the hood, the face velocity can increase to unsafe levels. At the top of a bypass hood there is another opening called a bypass. This bypass opens as the sash is closed and closes as the sash is opened. The total air coming into the hood between the sash opening and the bypass helps keep the face velocity at a safe level.

High Performance

The label "high performance" can be applied to a number of hood types. It is not really a class of its own. It generally means that the hood can perform safely at lower face velocities. Thus, achieving the same performance with less air. Hoods that carry this label are generally more robust, meaning that they have a better chance of performing well, even when the room conditions are not ideal. For example, if a high-performance hood tests well at 40 fpm under ideal conditions, it might do better than a standard hood at 80 fpm or 100 fpm when the overall laboratory ventilation system is weak.

VAV Restricted Bypass

This is a form of bypass hood intended for use with a VAV (Variable Air Volume) exhaust system. A VAV fume hood adjusts the volume of air automatically as the sash opens and closes. Because there is less air as the sash closes, the bypass is not needed for face velocity control, but because chemical build-up can occur within the hood when there is no air coming into the hood, the

An old joke from Engineering school is that if you put your head in the oven and your feet in a refrigerator, your average temperature would be just right. This illustrates the problem with using averages in some circumstances. One size fits all is not an applicable concept where fume hoods are concerned.

ASHRAE's Approach

To address the issue of risk management as it relates to chemical fume hoods, ASHRAE has created a new standard that is risk based. They looked at the levels of exposure and defined a level of protection. This coupled with the SEFA ECD guidelines has finally provided a workable risk-based concept.

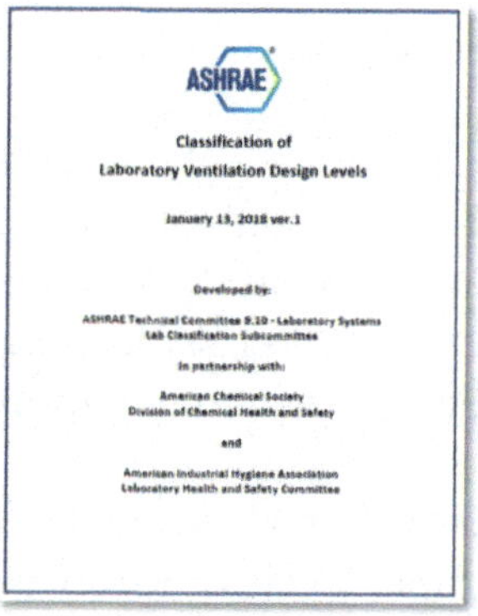
ASHRAE

Classification of
Laboratory Ventilation Design Levels

January 13, 2018 ver.1

Developed by:

ASHRAE Technical Committee 9.10 - Laboratory Systems
Lab Classification Subcommittee

In partnership with:

American Chemical Society
Division of Chemical Health and Safety

and

American Industrial Hygiene Association
Laboratory Health and Safety Committee

The use of risk-based design is long overdue and is a more logical way to balance safety and energy usage.

The ASHRAE Standard is called Classification of Laboratory Ventilation Design Levels. Much like the Biosafety levels, ASHRAE has established 5 ventilation system design levels. The levels are designated VSDL 0 through VSDL 4. The 5 levels correspond to risk levels from Negligible (0) to Extreme (4).

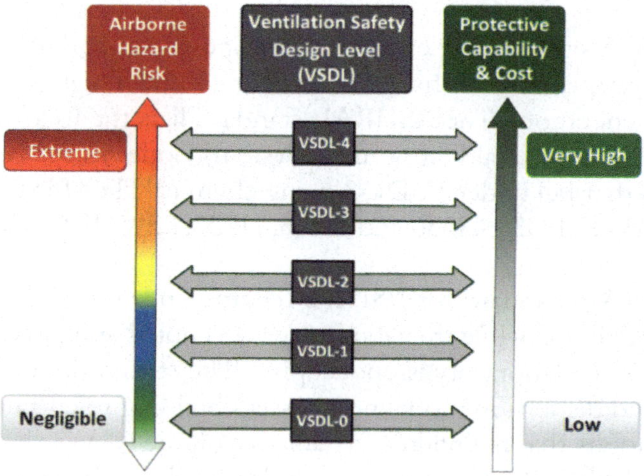

ASHRAE explains what types of chemicals and work are involved in each design level and gives performance criteria for each level.

While these classification standards can be implemented with fume hoods using today's technology, it offers some interesting future possibilities. As fume hood technology advances and fume hoods become more advanced, they could have the ability to adjust to different design levels. A fume hood could operate as a VSDL-2 this morning and switch to a VSDL-4 this afternoon.

This design level approach to fume hood performance seems to be the most adaptable to the current state of the industry and the technology currently in use. It also provides a platform for future innovation and the integration of "smart" ventilation systems.

For example, if the work I do in a certain laboratory is classified as VSDL-0, I really don't even need ventilation devices. The work can be done safely on the open countertop. However, if my work falls under the VSDL-4 classification, I must be much more concerned about safety and take many more precautions.

This ASHRAE safety level strategy gives us the opportunity to use multiple levels of safety precautions within the same building -- a zone concept. Instead of having too little protection in one area or unnecessary protection requirements in another, we can better match the risk of exposure to the level of protection we are providing.

In a VSDL-3 situation for example, the conditions are defined as having large quantities of hazardous materials and a high potential for airborne generation. The ASHRAE standard lists the following chemical GHS (Global Harmonized System) Codes as the types of chemicals that fall under VSDL-3 being chemical like H331, H332, H333, H334 & H335, H336 and H226, H227, 229, H242, H261.

The ASHRAE standard for VSDL-3 recommends 6 to 8 ACH (air changes per hour) while the laboratory is occupied and a reduction to 4 ACH when the laboratory is unoccupied. The standards additionally recommend dispersion modeling to assess building re-entrainment. Use of fume hoods that maintain constant air volume (CAV) should be considered when designing equipment for level 3 risks. In addition,

was happening around them. It was necessary to fence off the robots so that people and materials could not get in their path of operation. The cobot is a robot with self-awareness. It is able to sense its surroundings and adjust its behavior to fit its programing. It can avoid people or objects in its path.

A cobotic fume hood would possess the self-awareness necessary to provide operational data to the backend systems, so that a host of variables could be analyzed and performance of the entire space optimized. By humanizing the interface, the technology is more accepted by the user and the cobot is more likely to be considered a friend rather than a foe. A cobotic hood can become a virtual/digital laboratory assistant that helps the user be a responsible and more predictable component of the overall system. This new technology will have the ability to minimize danger and allow the user more time to focus on the science and less time on the routine tasks.
Today, there are a few laboratories where cobots work side by side with laboratory technicians. They perform many of the repetitive and boring tasks. At some point, these cobots will take over hazardous work. They will interface with the fume hood. To accomplish this there must be a method for the cobot to communicate with the fume hood.

Another automation coming to the laboratory is telepresence robots. These are remotely controlled, wheeled devices that use wireless Internet connectivity to communicate. Laboratory collaboration using telepresence is expected to become common. It also represents an opportunity for you to be in your lab without physically being there. Cobots are already using instruments and equipment and as more of these devices appear in the laboratory, there will need to be more communication between these digital devices.

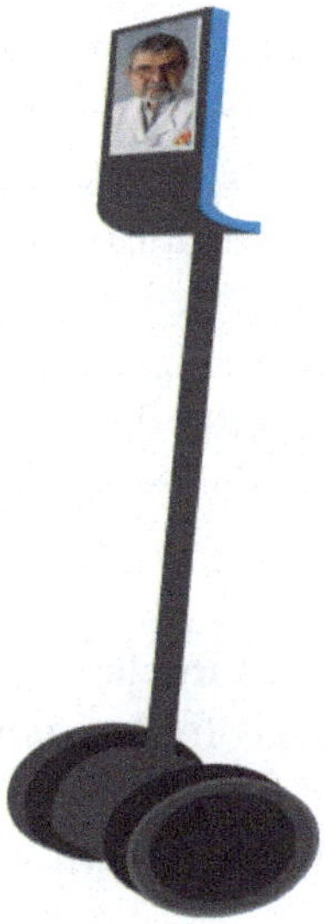

The future fume hood will have more of these digital features. In fact, it is likely that the term fume hood will be abandoned for a more accurate term. Perhaps a more descriptive term such as chemical containment device (CCD) will be adopted. Some manufacturers are already using this term. The CCD should assist and help the user make better decisions on how to use the device. While a CCD could assist multiple users, the interaction should be tailored to each person and what that person is doing at the time. As an aware device the CCD should consider what other activity is taking place in the laboratory. The performance variables of the fume hood can be adjusted based on risk.

What might an interaction with a cobotic fume hood look like?

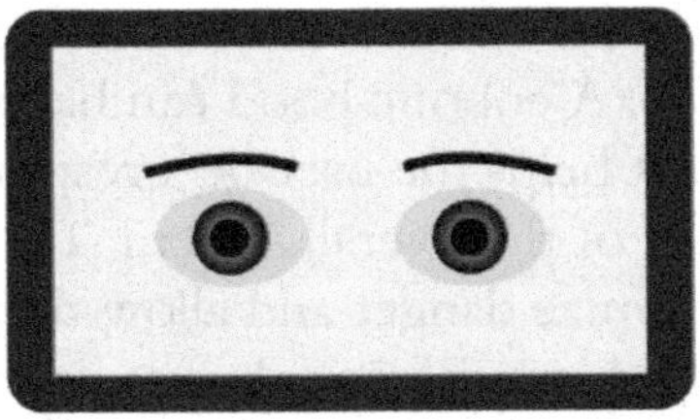

It all starts with a new interface. The concept is to create a digital partner that can assist the user in making good choices. The interface of a CCD will most likely be a tablet like interface that can be both a touchscreen that responds to a gloved hand and hand movements as well as voice commands. We want to humanize the interface as much as possible.

The CCD can provide a new level of security and control, making the experience safer and more productive. The first principle is access control. Access is controlled by a combination of radio-frequency identification (RFID) and/or Face/Voice recognition. So, the CCD recognizes the user and based on this recognition, the CCD can access information about the user such as training records. Knowing who is using the device is important to be able to personalize the experience. This data will also be useful in the back office to better understand usage.

In our intelligent laboratory, we want to humanize technology. The goal is for the devices to not only protect the individual, but to assist the person in making good choices and being more productive. The control system would have almost unlimited access to data.

The interface could be similar to Amazon's "Alexa" or Google Assistant's "Ok Google." To allow many people to work together in the same area, a unique action phrase or ID for each device would be used. In this example we will call our fume hood/CCD "Number 6."

The same technology could be expanded to other systems around the laboratory, much like the home automation hubs. The interface could be personalized to make the technology feel more comfortable or human.

"A future Smart Laboratory Interaction"

Assume that your name is Mark and you have named your digital assistant Abby. Abby is available on any intelligent device in the laboratory, in fact it could be accessible through any Bluetooth device.

You approach the locked door to the laboratory. The entry system is equipped with a thermal scanner and facial recognition. As you approach, the voice you have chosen for Abby says "Good Morning Mark." Or you could change your preference to provide you with a more personal or expanded greeting such as, "Welcome back Mark, remember you have a staff meeting at 10 a.m. this morning."

You respond, "Good Morning Abby" as you enter the laboratory. You proceed to the area where you work (your office, your work bench, your write-up area, or a specific bench area that has been assigned for the day), and when you are detected in that area, Abby would say, "Mark, what can I help you with?"

You might say: "What is on my schedule today?" or "What is the weather?" or "Is Joe Smith here yet?" or "Set reminder for ..." or "Schedule a meeting for 10 am with Joe Smith" or "Is Number 6 available?" This is a similar type of conversation that you might have with smart devices such as Amazon Alexa or Google Assistant.

Next you go into the laboratory and approach a fume hood (now called a CCD) where you intend to work. This CCD has been named "Number 6" (Each device has an individual name so several CCDs or smart devices can co-exist in the same space).

When you approach, Number 6 says, "Good Morning Mark, what are we doing today?"

You respond with your work plan.

Number 6 says, "Mark, in checking your files, I see that you haven't received current training for that process, can I show you a short training video?"

You respond, "Sure."

After the video training is complete, Number 6 asks you a few test questions to verify your understanding of the training material. You answer those questions. Then Number 6 says, "Mark, do you acknowledge that you understand the procedure and the risks?"

You say, "Yes I understand."

Number 6 says, "Ok, I will record in your file that you now have received the training."

Number 6 says, "This work requires the following personal protective equipment (PPE): latex gloves, safety goggles, and your laboratory coat. Please confirm you are using these items."

You say, "Yes I have my gloves and goggles and laboratory coat on."

Number 6 says, "I have calculated that the risk level for this procedure is a 4. I am adjusting the sash and airflow parameters to ensure your safety. Are you ready for setup?"

You say, "yes."

The sash opens and the lighting adjusts to a setup level. The CCD now shifts from unoccupied mode to setup mode.

A laser projects information on the worksurface showing the safe work areas. A setup checklist is presented on the display.

Number 6 asks, "Do you need assistance with the setup?"

You say, "Show me a typical setup for this task."

A photograph or diagram is shown on the display and a laser guideline is projected on the work surface. If this procedure/experiment has been done before, a photograph and other related information can be stored and displayed again. Photographs or reference materials can be called up by the user as needed, along with the Safety Data Sheet (SDS). At this point, the system knows who is using the CCD, what task will be performed and the conditions in and around the CCD. The internal camera takes photographs and video records the activity. The video loops every hour and only saves the footage based on preset conditions such as an accident or a red condition lasting more than a specified period of time. A photograph is taken every time the CCD changes mode and another each hour.
You perform the appropriate setup.

You say, "Ok Number 6, I am finished with my setup and am ready to start."

Number 6 says, "Would you like to see an airflow visualization?"

You say "Yes."

The sash adjusts to the operating position and the CCD switches to operational mode. The test fog starts coming out of test ports and you can see the impact of this setup on the airflow. After a minute, Number 6 asks, "Are you satisfied with the setup?"

You say "Yes." The fog stops and you are ready to start.

Number 6 asks, "Would you like me to send you notifications about this experiment as it progresses?"

You say, "Yes."

You go back to your office and monitor the experiment via video receiving real-time data and updates on the progress of your project.

While you are away, the monitor on the fume hood displays what procedure is being performed along with your contact information. If the CCD was outfitted with a cobotic arm, some of the steps normally performed by a human could be automated.

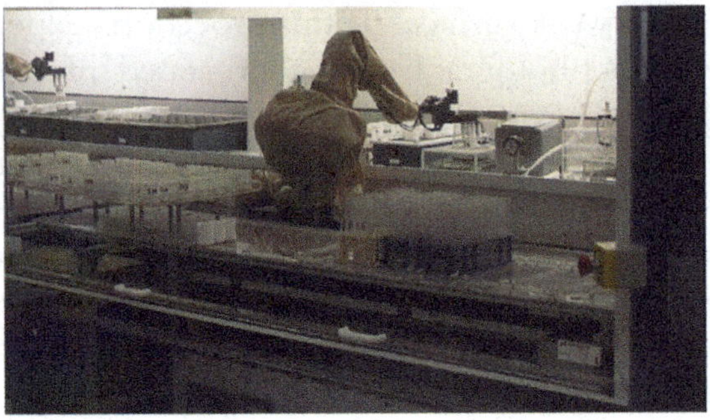

The CCD knows what is happening on the inside of the fume chamber as well as within the laboratory room in general. An exposure risk factor has been calculated and CCD performance has adjusted accordingly. The CCD is logging data from all sensors. Because many laboratories will have multiple CCDs operating at the same time, the combined data is shared in an Internet of Things (IoT) style, allowing more interaction of devices and better predictions.

At the CCD, the display is a graphic that shows what is happening inside the hood and the background is either, green, yellow, or red based on the predictive containment calculations -- Green indicating good containment, yellow indicating caution, or red indicating the danger of containment loss. The display shows the user's name and contact information. It also has an emergency button. It might additionally display elapsed time, or time remaining, or other critical data.

This display is visible to others in the laboratory so that they can see who is using the CCD, what is being performed, and the predicted containment level along with other useful information.
Assuming that out of a 7-hour day, the CCD is not used for 2 hours, and is in setup mode for 1 hour and only working 4 hours, the CCD can automatically adjust from standby mode, to setup mode, to operating mode for a level 4 risk. In those 7 hours the CCD might have used 50% less energy than used today in a typical system and the

laboratory will be safer. Because the CCD is communicating with the other CCDs and the users, it can anticipate system changes, minimizing system response time and greatly improving the probability that the CCD will remain in a green condition.

The collection of sensor data and user inputs is a large amount of data that is sent to a data lake for storage. The amount of data collected makes the CCD a "Big Data" device. Using Big Data techniques, the data can be mined and reported in an easy to use dashboard.

Examples of situations in which this data could be useful include:

A particular CCD has been showing a high number of abnormal red conditions. The data shows that a particular user is present for most of the events. By reviewing the setup photographs and other related data it can be determined whether this problem is related to a training issue that can be addressed with the user or whether it is a malfunction of equipment.

A particular laboratory room is showing a high number of red conditions from multiple CCDs. Analysis of the data might suggest a balance problem or equipment issues that can be addressed by maintenance.

This data can provide a more holistic view of the laboratory and its operation. The data can be trended to show user patterns and requirements for services. With this data platform, the laboratory can be operated at a higher level of safety, reducing risks, reducing energy usage, and generally increasing efficiency and productivity.

This is just a quick overview of how a smart CCD might behave and interact with a person. The introduction of artificial intelligence (AI) and machine learning will change the way we think about laboratory space and how we manage it.

While not all the functions will be present in the first generation of smart fume hoods, they are all realistic possibilities as the technology in laboratories evolves. While it may take several generations of products to advance to this level, we can expect continual progress. Much like smartphones where we see a new generation every year, each new model of CCD will have more features and capabilities.

It will probably take a decade for the concept to mature, but if you can see a future, you can make it happen. Our goal is identifying the problems we are addressing and then look for innovative solutions. The smart hood will be like many other smart devices. Different manufacturers will introduce different features in different generations of the product. Like smart phones and smart devices, designers of smart CCDs will learn from the market what features are most effective and desired. Innovation and progress will build on the successes and each generation of product will improve.

Again, for the fume hood/CCD to function safely, it needs to know what is happening in and around it as well as how the laboratory ventilation system and the various building systems are performing. Instead of being just a component in the mechanical system, it will become smart and an active participant in the safe operation of the laboratory.

Over the next decade these are some of the features we could see in fume hoods of the future:

- These intelligent CCD devices will be outfitted with various sensors to monitor various conditions and predict performance. There is nothing intelligent about today's fume hoods, they are just boxes that exhaust. Much like a robot, they are totally unaware of anything around them. They can perform as programmed, but the performance is mostly static and has little to do with the risks or with what the fume hood is actually being used for. Thus, we are dependent on the user to be smart and use the hood safely. But even a safe user has minimal control over the system performance.

- The concept of a smart CCD is to be self-aware much like a cobot. The CCD needs to be able to sense what is going on inside the hood and around it so that the risk of exposure to the users can be assessed. The ability to communicate with the laboratory ventilation system is important so that the CCD can manage the operational parameters to maximize safety and minimize energy usage.

to use the CCD and whether the user has training specific to the procedure to be performed. The smart CCD can offer point of use training and safety reminders.

- Differential pressure between the inside of the fume hood and the laboratory is one of the best indications of momentary loss of containment. Strong negative pressure within the fume hood is a good indicator of containment when factored with other variables

- Room Temperature and humidity impact the air movement. Temperature gradients within the room impact turbulence and cross drafts. Monitoring and understanding these factors can help improve CCD performance.

- The temperature inside a fume hood is a micro version of what is happening in the laboratory. Monitoring this temperature helps predict turbulence and potential problems associated with the process, such as excessive heat or fire.

- Automatic sash movement will eliminate a safety risk. The sash not only becomes a physical safety barrier, it becomes a means of access control. Remembering to lower the sash to the operating position or to close it when finished becomes a thing of the past.

- It will be possible for the CCD to maintain a usage log automatically and to summarize any level of usage detail. Information such as who used the CCD, for how long and how many red conditions occurred during their usage will be gathered for later analysis.

- The ability to generate fog will allow the visualization of airflow within the hood and show the impact of the equipment setup on airflow. This tool allows the user to make adjustments to the setup before loss of containment occurs.

- Laser projection of images on the worktop, from simple boundaries to complex setups, can assist the user in fast and proper setup.

- Because the CCD is voice controlled, users will be able to use voice controls to record information and data. Voice controls can also be used to provide information, record notes and log data.

- Future CCDs will interface with building systems (general HVAC, laboratory ventilation, fire, security).

- The ability to integrate augmented reality (AR) and virtual reality (VR) into the CCD setup and procedures brings about a whole new experience.

These are just some of the features that can be performed by a smart CCD. Until these smart units are put into use, we will not know what features are most desired and effective, but the trend to use smart devices has already started. It is clear that the fume hood as we know it today is problematic at best and as the technology improves innovators and designers will use new technology to go outside the box to find better solutions to the problems. For the first time in a hundred years, the next generation of smart devices are within reach. Finally, fume hoods are poised to live up to their intended use of providing maximum safety to users while being environmentally sensitive and minimizing energy usage.

The year 2020 and COVID-19 have brought about many changes in how we view and use technology and how we work. The digital transformation has advanced at least 5 years in a single year. The impact of how we will work during a pandemic will impact laboratories for many years to come. As a group, I think we will be more open to technology as a way to more effectively work in the future. The items discussed above are not science fiction, many are being developed today or are already in the market. The future is now.

Chapter 15

The History of Fume Hoods

While working at Kewaunee Scientific during the 1980s, I was given a project to research the company's history and the history of the fume hood industry. I found the project intriguing for a number of reasons. First, it was a very interesting chain of events that led to the current status of the fume hood market. Equally enlightening was the realization of how much can be learned from the past. The reasons behind current fume hood practices and the analysis of past failures and successes provide valuable lessons. To preserve this legacy, I am ending this book with the history of fume hoods. I hope you find it as interesting as I do.

Timeline for the development of fume hoods.
The need for ventilation has been apparent since the early days of chemical research. The following is a chronology of important milestones in fume hood development.

600 AD to 1600 AD -- The fume hoods of this period were fireplaces used by alchemist in the middle ages. The origin of the modern fume hood lies in the fireplaces of ancient alchemists who practiced a combination of chemistry, physics, metallurgy, and mysticism, often in attempts to convert base materials into precious metals.

1790 -- Joseph Priestley designed a chemical exhaust system for his laboratory in Pennsylvania operated by large manpowered bellows.

Early 1900s -- One of the earliest scientists concerned with laboratory ventilation was Thomas Edison who used the fireplace chimney to exhaust noxious fumes and odors from his laboratory. Thomas Edison used a fireplace as a fume hood on cold winter days, but on warmer days Edison would place a shelf outside his laboratory's double hung vertical window. This was the genesis of today's fume hoods.

1923 -- The first recognizable fume hood, in the modern sense of the word, was developed at the University of Leeds. It consisted of a large cabinet standing at working height and incorporated vertical rising sashes arranged like parallel windows. It was a takeoff on Edison's fume hood and was designed by interior decorators as a cupboard using vertical counter weighted sash widows. This device was referred to as a fume cupboard. To this day, in Europe, fume hoods are called fume cupboards. This laboratory style furniture is why many consider the fume hood a furniture item rather than the mechanical device it is.

1939 - 1945 -- During World War II considerable advances were made to fume hood technology in response to fears of toxic chemical exposure and radiation. This resulted in improvements in design, safety and ventilation. The first real fume hood design changes occurred due to the numerous deaths of researchers working on the atomic bomb project. Designers learned anecdotally that no particular face velocity made a fume hood safe, but they never understood why. It took another 50 years to learn the reason why fume hood face velocity alone did not make a laboratory safe.

Also, during the war years, the high-efficiency particulate air (HEPA) filter was developed by a group which later became the Atomic Energy Commission. The development of the HEPA filter had a dramatic impact on the effectiveness of fume hoods and biological safety cabinets, greatly increasing protection for users.

1943 – John Weber, Jr. working at the Ames Laboratory in Ames, Iowa, developed the concept of a constant face velocity, variable exhaust flow fume hood control. This design was applied to a vertical rising sash hood served by a dedicated hood exhaust fan. The concept eventually became a standard feature employed on many fume hoods. At that time, it was used in atomic laboratories, especially where ventilation containment within the hood was critical.

1951 – H.W. Alyea, chief field engineer at the Johnson Service Company (now Johnson Controls, Inc.), realized that keeping the door of a fume hood closed as much as possible, and certainly when not in use, resulted in a considerable decrease in the amount of airflow through the laboratory. Along with this reduction in the air supply air came proportional reduction in cooling demand and considerable energy savings

Early 1950s -- John Turner, working in the Engineering Department at Oak Ridge National Laboratory (ORNL), suggested replacing vertical rising sashes with horizontal sliding sashes in order to reduce energy consumption. He also introduced the use of a mechanical damper that moved as a result of the imbalance between external and internal hood pressures.

Around the same time, Weber also recognized that the best containment in a fume hood was achieved with a minimum sash opening. This awareness resulted as he developed the emergency quick close feature incorporated as a part of his system. The exhaust fans used as a part of Weber's system were never turned off.

By 1960, fume hoods had been established as box-like enclosures with a movable sash. The baffle systems resembled a traditional fireplace and the laboratories they were installed in had pressurized air supply. The exhaust system was never turned off.

The 1960s -- Many suppliers joined the market with different concepts and many products that didn't really work that well. The lack of product standards allowed a wide range of products to come to market. 1961 – Labconco introduced its first one-piece molded fiberglass-lined fume hood. Fiberglass was chosen to line the fume hood because of its durability, cleanability, high light reflectivity, fire resistance and chemical resistance properties.

1964 – Kewaunee produced the Aristocrat Auxiliary Air Fume hood. Kewaunee's work with the government on the Manhattan Project made them a leader in fume hood development. The introduction of auxiliary air fume hoods conserved energy by introducing outside air in front of the hood. This reduced the loss of tempered air from the laboratory. This type of fume hood required the use of two duct and blower systems.

1968 – François-Pierre Hauville created the company Erlab and began selling the first ductless fume hood the same year.

1970s -- Fume hoods constructed from wood were replaced with epoxy painted steel. During this time, the walk-in hood was introduced, as well as a number of special purpose hoods such as perchloric acid and radioisotope hoods.

1980s -- The industry responds to a market flooded with poor products. The writing of fume hood standards began and ASHRAE 110 -1985 was released for testing hoods. As ASHRAE 110 was being specified, companies were required to either improve the performance of their products or exit the market. The end result was that many of the unsafe products left the market. During this time, the precursor to SEFA (SAMA) wrote a fume hood standard.

More changes in fume hood design and performance occurred in the 1980s than in the previous 50 years. Most of the lessons learned were by trial and error, but a full understanding of the laboratory ventilation system and the factors that impact a fume hood's performance were still elusive. Another major change that occurred during this time was the elimination of liner materials that contained asbestos. During this same period there was a huge improvement in testing equipment that made it possible to do containment testing.

1985 -- AFNOR introduces the NFX 15-211 ductless filtering fume hood performance safety standard.
Variable Air Volume (VAV) technology was being applied to fume hoods to conserve energy.

1990s -- Several major improvements to fume hood technology occurred. New material technologies, and requirements for better chemical and flame resistance lead to a number of new design concepts. Flush sills were introduced as well as auto sashes and chain driven sash systems.

1991 – The Occupational Safety and Health Administration (OSHA) was given jurisdiction over laboratories, which had been previously excluded. This government standard OSHA 29CR-1910.1450 was the first government standard that addressed fume hoods.

1993 – Tom Saunders wrote the book "Laboratory Fume Hoods – A User's Manual" in an effort to explain to users how the fume hood was intended to be used.

1994 – SEFA 1 for fume hoods was written. This was an industry standard that updated the previous SAMA fume hood standard. Also, the CSA Z316.5 ductless fume hood safety standard was written.

1995 – A new version of ASHRAE 110 was released that greatly improved the quality of fume hood containment testing.

1996 – The axle, chain and sprocket sash system was developed by Chip Albright.

1998 – Computational Fluid Dynamics (CFD) modeling began to be applied to fume hoods.

1999- Underwriter's Laboratory released UL1805 which provided requirements for laboratory hoods and cabinets addressing many of the safety issues that were being recognized in the market. Additionally, a greatly expanded SEFA 1 Fume Hood Standard was written.

During the 1990s, variable air volume (VAV) hoods began to outnumber constant air velocity (CAV) hoods, especially in laboratories that were 24/7 systems.

2000s -- Driven by demand for more energy efficient models, low flow fume hoods, operating at face velocities of 40-80 fpm, were developed. These fume hoods delivered excellent performance while saving energy and money. However, the concept of "safety first" was overshadowed by concepts that saved energy and money and didn't always provide optimal safety.

2008 -- ERLAB introduced the filtering fume hood technology also known as GreenFumeHood which incorporated Neutrodine multistage general-purpose filtration for use in ductless fume hoods.

2012 – The ANSI Z9.5 standard was updated.

2016 – The ASHRAE 110 standard was updated.

2018 – The concept of synchronized supply brought a new class of fume hoods to the market.

2019 – The EN 14175 standard was updated.

Now we are up to current times, I hope you have learned more about fume hoods and what is necessary for them to provide the intended safety. As I said in the previous chapter, I am very excited about the future of laboratory ventilation and chemical containment devices. Smart hoods are coming and most likely from Asia, but as this history is updated, I see this as a defining moment for fume hoods and laboratory safety.

As I conclude this book, I can honestly say that I am optimist about the future. As more people understand the issues at play and we begin to address core problems rather than symptoms, technology will help us displace many of the bad habits and unsafe practices. User safety will improve and we will be able to make labs greener. Technology will allow us to work smarter and be more productive. Much of the routine and boring tasks can be automated allowing the scientific minds to be more creative and experience greater discovery.

Thank you for taking the time to read this book and I hope you have found it worthwhile. I feel the best way to improve is through education. Hopefully this book will help make you a safer hood user.

Glossary of Terms

These are terms that are used in this book and are common when discussing fume hoods and laboratory ventilation systems:

Acceptable Indoor Air Quality: Air in which there are no known contaminants at harmful levels as determined by appropriate authorities and air which 80% or more of the people find acceptable.

Access Opening: The part of the fume hood through which work is performed; sash or face opening.

ACH, AC/H (air changes per hour), N: The number of times air is theoretically replaced in a space during an hour. An ACH rate for a room can be converted to volumetric airflow by multiplying the ACH number times the gross volume of the room.

Adjacent Roof Line: For the purposes of determining the laboratory chemical hood stack height, the adjacent roof will be within 6 feet horizontally of the nearest exhaust fan stack.

AHU, Air Handler Unit: A device used to regulate and circulate air as part of a heating, ventilating, and air-conditioning (HVAC) system. An air handler is usually a large metal box containing a blower, heating or cooling elements, filter racks or chambers, sound attenuators, and dampers. Air handlers usually connect to a ductwork ventilation system that distributes the conditioned air through the building and returns it to the AHU. Sometimes AHUs provide (supply) and or they can accept (exhaust).

Airflow Monitor: Device installed in a fume hood to monitor the average airflow through the sash opening.

Air Foil: Curved or angular member(s) at the fume hood entrance. The lower airfoil is a horizontal member across the lower part of the sash opening to provide a smooth air flow into the fume chamber across the work surface into the baffles.

Air Lock: An intermediate chamber between two dissimilar spaces with airtight doors or openings to each of the spaces. The doors are interlocked to ensure that at least one of them is always closed.

Air Volume: Quantity of air expressed in cubic feet (ft3) or cubic meters (m3).

Artificial Intelligence (AI) and Machine Learning: Artificial intelligence (AI), machine learning and deep learning are three terms often used interchangeably to describe software that behaves intelligently. This software can evaluate vast amounts of data and make intelligent decisions.

Auto Sash: Is a feature where a fume hood sash has been motorized and can be controlled without using one's hands.

Auxiliary Air: Supply or supplemental air delivered to a laboratory fume hood to reduce room air consumption.

Auxiliary Air Hood: A laboratory chemical hood with an external supply air plenum at the top of the laboratory chemical hood. The auxiliary air plenum provides a makeup airstream comprised of unconditioned or only minimally conditioned outside air to substantially reduce the amount of conditioned room air exhausted by the laboratory hood.

Baffle: A panel located across the rear wall of the fume hood that directs the airflow through the fume chamber. The baffle creates a plenum at the rear of the hood to direct air to the exhaust outlet.

BAS, Building Automation System: Is an intelligent system of both hardware and software, allowing the heating, venting and air conditioning system (HVAC), lighting, security, and other systems to communicate on a single platform.

Bench Hood: A fume hood that is located on a counter height work surface.

Biological Safety Cabinets, BSC: A fume hood like device that is intended for use with biological agents and not for chemical use.

Blower: see Fan.

Blower Exhaust Velocity: The speed at which air is exhausted from the fume hood exhaust stack. The minimum should be at least 3000 fpm (15.2 m/s).

Bypass: Compensating opening in a fume hood that functions to limit the maximum face velocity as the sash is lowered.

Bypass Hood (constant air volume bypass laboratory hood): A laboratory hood design that incorporates an opening (bypass area) in the upper portion of the laboratory hood structure. When the movable sash is fully open, the sash blocks off this bypass area and all of the airflow into the laboratory hood must pass through the open face area. However, as the sash is being closed to reduce the open face area, at a specific point an amount of bypass area is being uncovered. The increase in the bypass area opening offsets the decrease in the face area opening, thus providing an alternate path (the uncovered bypass area) for air to flow into the laboratory hood. When utilized with a constant air volume ventilation system, the bypass area keeps the laboratory hood face velocity relatively constant and from increasing to an objectionably high value as the sash is lowered.

Canopy Hood: Ventilating enclosure suspended above work area to exhaust heat, vapor or odors. This device is not a laboratory fume hood.

Capture Velocity: Speed of air flowing past the face opening through a fume chamber at a speed necessary to capture generated fume vapors and/ or particulates and directed to the exhaust outlet. Measured in feet per minute (fpm) or meter per second (mps).

CCD Chemical Containment Device: A new class of Exposure Control Device that represents the Next Generation of Laboratory Fume Hoods. These devices incorporate new technology not currently present in traditional fume hoods.

CFD (Computational Fluid Dynamics): Is a branch of fluid mechanics that uses numerical analysis and data structures to analyze and solve problems that involve air flows. Computers are used to perform the calculations required to simulate and visualize the free-stream flow of the air within surfaces defined by boundary conditions. With high-speed supercomputers, better solutions can be achieved. A good tool for fume hood design, but generally lacking the definition to truly understand how the air is behaving.

Chemical hygiene officer: An employee who is designated by the employer and who is qualified by training or experience to provide technical guidance in the development and implementation of the provisions of the Chemical Hygiene Plan.

CHP (Chemical Hygiene Plan): Is a written program developed and implemented by the employer which sets forth procedures, equipment, personal protective equipment, and work practices that are capable of protecting employees from the health hazards presented by hazardous chemicals used in a particular laboratory.

Clean Air Hoods: These fume hood like devices are designed to provide product protection. Often found in cleanrooms, they are not intended for user protection against exposure.

Cobotic Hood: This is future generation of fume hood or chemical containment device that is driven by AI and actively participates and assists the user with safe operations.

Color Temperature: Color temperature is conventionally expressed in kelvins, using the symbol K, a unit of measure for absolute temperature. Color temperatures over 5000 K are called "cool colors" (bluish), while lower color temperatures (2700–3000 K) are called "warm colors" (yellowish).

Combination Hood: A fume hood assembly containing a bench hood section and a floor mounted section.

Combination Sash: A fume hood sash with a vertical framed member that moves vertically for setup and houses two or more horizontal sliding transparent viewing panels that can be opened for operation.

Constant air volume (CAV) ventilation system: A ventilation system designed to maintain a constant quantity of airflow within its ductwork. Although relatively simple, a constant volume ventilation system typically requires the maximum ongoing energy usage since the system always operates at maximum capacity.

Counter Top: (See Work surface)

Cross Drafts: Air drafts or currents that flow parallel to or across the face opening of the fume hood.

Damper: Device installed in a duct to control airflow volume.

Data acquisition, DAQ: Is the process of measuring an electrical or physical phenomenon such as voltage, current, temperature, pressure, or sound with a computer. A DAQ system consists of sensors, DAQ measurement hardware, and a computer with programmable software.

Data Lake: Is a centralized repository that allows for storage of structured and unstructured data at any scale without having to first structure the data. Using dashboards for visualization and real-time analytics it supports machine learning and is a guide to better decision making.

Demonstration Hood: A vented enclosure used for student demonstrations that has visibility on at least three sides, used primarily in schools. This device is not a laboratory fume hood.

Design Sash Position: The maximum open area of the hood that achieves the desired face velocity.

Differential Pressure (DP): The difference in pressure between two points of a system, such as between a room and the adjoining hallway, or the fume chamber and the room.

Dilution Ventilation: Ventilation airflow that dilutes contaminant concentrations by mixing with contaminated air, as distinguished from capturing the contaminated air.

Discharge Velocity: The speed of the exhaust air normally expressed in feet per minute (meters/second) at the point of discharge from a laboratory exhaust system. It is the air velocity as it leaves the last element of the exhaust system. The term "stack velocity" is sometimes used when referring to the speed of the exhaust airstream as it is discharged into the outside air.

Distillation Hood: A laboratory fume hood that provides a work surface approximately 18 inches (45.7 cm) above the room floor, to accommodate tall apparatus.

Diversity: Operating a system at less capacity than the sum of peak demand to reduce system size and energy usage.

Diversity Factor: A percentage factor that is applied to establish the theoretical maximum exhaust airflow quantity that is required at any point in time. For example, in an exhaust system consisting of six hoods, it could be assumed that no more than three hoods would be operated at one time. A diversity factor of 50% could be applied. Applying a diversity factor to the theoretical maximum required capacity of an HVAC system is often considered in the design of a VAV system. Incorporating a diversity factor enables downsizing HVAC system components and thus results in a smaller capacity ventilation system. The overall intention of applying a diversity factor when designing a VAV ventilation system is to achieve a lower life cycle cost (e.g., lower system first cost and/or lower system energy costs).

Dual Entry Hood: A bench type fume hood that has two sash openings, usually on opposite sides. These are often used in education and are not a laboratory fume hood.

Duct: Round, square or rectangular tube used to enclose moving air.

Duct Collar: Is the point where the fume hood connects to the duct.

Duct Transition: Is a section of duct that connects two dissimilar shapes or sizes of duct. For example, a rectangular duct to a round duct.

Duct Velocity: Speed or air moving in a duct, usually expressed in feet per minute (fpm) or meters per seconds (mps).

Ductless Hood: A fume hood that is not connected to a laboratory ventilation system, rather, a ductless hood incorporates an exhaust fan and exhaust filters as an integral part of the design and discharges the exhaust directly back into the room. Ductless laboratory hoods are of limited size and capacity in comparison to conventional ducted laboratory hoods. These are for special applications and care must be taken when choosing the correct filter.

ECD, Exposure Control Device: This is any class of product that is designed to minimize the exposure of users to hazardous chemicals.

EHS, Environmental Health and Safety: This is an organizational function whose responsibility is to inspect and monitor environment, machineries and processes to ensure safety as per government rules and regulations and industry standards. They have knowledge of government rules and regulations and prepare standard operating procedures that helps users be safe and healthy.

Electrical Service Fixture: Outlet or other electrical device mounted directly to the face of the fume hood.

Exhaust Air: Air that is removed from an enclosed space and discharged to atmosphere.

Exhaust Collar: Connection between duct and fume hood through which all exhaust air passes.

Exhaust Plume: This is the stream of contaminated air being discharged for the exhaust stack.

Exhaust Stack: Is the tube connected to the output of the exhaust fan. It should have a minimal height of 10 feet or 3 meters above the roof line.

Exhaust Unit: Air moving device, sometimes called a fan or blower, consisting of a motor, impeller and housing.

Face: Front access or sash opening of a laboratory fume hood.

Face Opening: measured in width and height.

Face Velocity: Average speed of air flowing perpendicular to the face opening and into the fume chamber of the fume hood and expressed in feet per minute (fpm) or meters per second (MPS), measured at the plane of the sash.

Face Velocity Monitor: This is a device mounted on the face of the fume hood and displays average face velocity. It often has an alarm when the velocity drops below the programmed minimum.

Facial Recognition: A biometric technology capable of identifying or verifying a person from a digital image.

Fan: Air moving device, can be called an exhaust unit, or a blower, consisting of a motor, impeller and housing.

Fan Curve: A curve relating pressure vs. volume flow rate of a given fan at a fixed fan speed (rpm). This is necessary to properly size a fan.

Filter: Device to remove particles from air.

Friction Loss: The static pressure loss in a system due to friction between moving air and the duct wall; expressed as inches w. g. 100 feet, or fractions of VP per 100 feet of duct.

Flame Resistance: The ability to withstand flame.

Floor-mounted hood (walk-in hood): A larger-size laboratory hood with sash and/or door arrangement that enables access from the floor to the top of the hood interior. The name unfortunately is a misnomer and although the design and height of these hoods may allow it, users should not walk into any hood that may represent a significant exposure hazard. Walk-in laboratory hoods enable larger equipment and apparatus (e.g., equipment on carts, gas cylinders, etc.) to be more readily put in and set up within the laboratory hood.

Fog machine: A machine that generates theatrical fog for airflow visualization in a fume hood.

Fume Chamber: The interior of the fume hood measured in width, depth, and height.

Fume Cupboard: European term for laboratory fume hood.

Fume Hood Controller: Is a device that monitors different elements of hood performance and function. It can control the sash, lights, interface display and VAV for exhaust and supply.

Fume Removal System or Laboratory Ventilation System: Is engineered to effectively move air and fumes consistently through fume hood, duct and exhaust blower. **Note:** Room air, make-up air, auxiliary air (if used) and pollution-abating devices (if used) are integral parts of a properly functioning system and should be considered when designing a fume removal system.

Fume Removal System: A fume hood exhaust engineered to effectively move air and fumes consistently through fume hood, duct and exhaust unit.

Gauge Pressure: The difference between two absolute pressures, one of which is usually atmospheric pressure; mainly measured in inches water gauge (in. w. g.).

GEX, General Exhaust: General exhaust is used to keep the room in balance. Based on the required ACH, if there is not enough air being exhausted through fume hoods and other exhaust devices, air is exhausted through the general exhaust.

Glove Box: A totally enclosed and controlled environment work area providing a primary barrier to confine and contain hazardous materials within. This device is not a laboratory fume hood.

Grille: A louvered or perforated face over an opening in an HVAC system.

Hazardous chemical: any chemical that is likely to be harmful to human health. Often defined by Government regulations and industry best practices.

Heat Resistance: The ability to withstand heat without deteriorating.

Health Hazard: Any chemical that is classified as posing one of the following hazardous effects: acute toxicity (any route of exposure); skin corrosion or irritation; serious eye damage or eye irritation; respiratory or skin sensitization; germ cell mutagenicity; carcinogenicity; reproductive toxicity; specific target organ toxicity (single or repeated exposure); aspiration hazard or simple asphyxiant.

HEPA: High Efficiency Particulate Air (filter) Common standards require that a HEPA air filter must remove—from the air that passes through—at least 99.95% (European Standard) or 99.97% (ASME, U.S. DOE) of particles whose diameter is equal to 0.3 μm.

High Density Shielding: A barrier made of lead. Often used in Radioisotope Fume Hoods

High Heat Hoods: Heat loads in the fume chamber of a hood impacts both the airflow and containment. Some hoods are designed especially for working with high heat loads such as high temperature acid digestion.

Hood: A device which encloses, captures, or receives emitted contaminants. A hood and a fume hood are not the same thing. Often times a hood is vented enclosure.

Hood Entry Loss: The static pressure loss, stated in inches on a water gauge when air enters a duct through a hood. The majority of the loss is usually associated with a vena contracta formed in the duct.

Hood Static Pressure: The sum of the duct velocity pressure and the hood entry loss; it is the static pressure required to accelerate air at rest outside the hood into the duct at duct velocity.

HVAC: Heating Ventilating and Air Conditioning. Ventilation systems designed primarily for temperature, humidity, odor control, and air quality.

Imbalance: Condition in which ratio of quantities of supply air is greater or lesser than the exhaust air.

Inches of Water (inch w.g.): The pressure exerted by a column of water one inch in height at a defined reference condition such as 39°F or 4°C and the standard acceleration of gravity.

Indoor Air Quality, IAQ): indoor air quality related to temperature, humidity, CO_2 and airborne contaminants.

IoT (Internet of Things): Is a system of interrelated computing devices, mechanical and digital machines provided with unique identifiers (UIDs) and the ability to transfer data over a network without requiring human-to-human or human-to-computer interaction. The ability to communicate with other devices.

IPA, Isopropyl Alcohol: Is being evaluated as a new way to do fume hood containment testing. It may replace SF_6 as the standard for containment testing. The IPA would be atomized inside the fume hood and its escape measured.

Isolator: A core component to the pharmaceutical industry, critical for a range of processes. These gas-tight enclosures provide a complete barrier to ensure aseptic conditions and containment. They are a class of ECD, exposure control devices.

Laboratory: A building, part of a building, or other place equipped to conduct scientific experiments, tests, investigations, etc., or to manufacture chemicals, medicines, or the like. A place to conduct experimentation, investigation, observation in support of research and discovery. An area in which diverse mechanical services and special ventilation systems are available to control emissions and exposures from chemicals and harmful substances.

Laboratory Air Quality: Unlike normal indoor air quality, laboratory air is prone to have other chemical contaminants. The nature of the work in the laboratory has the potential to release these contaminants into the air. Without proper ventilation, the concentration of chemical contaminants can present health hazards.

Laboratory Fume Hood: A Laboratory Fume Hood shall be made primarily from flame resistant materials including the top, three fixed sides, and a single face opening. Face opening is equipped with a sash

Polyvinyl Chloride (PVC): A water insoluble, thermoplastic resin derived by the polymerization of vinyl chloride used chiefly for containers, fittings and piping.

Polyvinylidene Fluoride (PVDF): Material is a strong and abrasion resistant fluoropolymer. It is chemically resistant to most acids, bases and organic solvents, and is the preferred material for piping and faucets for ultra-pure water. Pure PVDF is an opaque white resin that is resistant to UV radiation, and is superior for non-contaminating applications.

Positive Air Pressure: Air pressure higher than ambient.

Posts (columns, facia, pillars): These are the vertical members of the fume hood face. Generally, they form an angular or radiused entry into the fume hood. They are also used for mounting the electrical and service fittings. They work with the upper and lower airfoils to direct air into the hood in a nonturbulent way.

Predictive Containment: Is a forecasting system that uses sensor output and processes it to predict the likelihood that the fume hood is safe and has no loss of containment.

Pressure Drop: The loss of static pressure between two points; for example, "The pressure drop across an orifice is 2.0 inches w.g."

Radioisotope Hood: Provides protection from applications requiring the use of radiochemicals. It has a fully sealed integral work surface reinforced to support lead shielding and coved interiors to facilitate decontamination.

Recirculation: Air removed or exhausted from a building area and ducted back to an air-handling system where it is mixed with outside fresh air. This air mixture is then conditioned and utilized for ventilation. Since air removed from a space is more often close to the temperature and humidity of the building interior than the outside air, the recirculation process enables achieving a greater reduction in heating and cooling energy than once used air where 100% outside air was utilized.

Reentrainment or reentry: The flow of contaminated air that has been exhausted from a building is drawn back into the building through air intakes or openings in the walls or doors and windows.

Register: A combination grille and damper assembly. These supply air outlets can also be called diffusers or grilles.

Relative Humidity (RH): The ratio of water vapor in air to the amount of water vapor air can hold at saturation. A "RH" of 100% is about 2.5% water vapor in air, by volume.

Remote Control Valves: Valves usually installed in the walls of fume hoods with the control handles normally on the face of the hood which regulate and control the flow of the services to the outlets in the interior of the fume hood.

Replacement Air: (Also, compensating air, make-up air) Air supplied to a space to replace exhausted air and maintain balance.

Respirable Particles: A collective group of fine solid particles, aerosols, mist, smoke, dust, fibers and fumes are called Respirable particulates.

Return Air: Air which is returned from the primary space to the AHU for recirculation.

Reynolds number (Re): Is an important dimensionless quantity in fluid mechanics used to help predict flow patterns in different fluid flow situations. The Reynolds number is used to determine whether a fluid is in laminar or turbulent flow.

RFID, Radio-frequency identification: Uses electromagnetic fields to automatically identify and track tags attached to objects. An RFID tag consists of a tiny radio transponder; a radio receiver and transmitter. When triggered by an electromagnetic interrogation pulse from a nearby RFID reader device, the tag transmits digital data, usually an identifying number, back to the reader.

Room Air: That portion of the exhaust air taken directly from the room.

Room Air Balance: A general term describing the requirement that a laboratory room have the proper relationship with respect to the total exhaust airflow from the room and the supply makeup airflow. The relationship of these airflows also establishes the pressure differential between the laboratory room and adjacent rooms and spaces.

Room Balance: In a laboratory we usually want the room to be under slight negative pressure. That means we balance the room to have slightly less supply air coming into the room in relationship to the air being exhausted.

Room Controller: This is an electronic device that processes data from various inputs and then communicates with the Laboratory Ventilation System to manage the exhaust and supply air to maintain room balance.

Room Ventilation: The volumetric airflow through a room expressed in terms of cfm or L/sec.

Sash: A moveable panel or door set in the access opening of a fume hood to provide access, and to provide a protective shield.

SCFM (Standard Cubic Feet Per Minute): Airflow rate at standard conditions; dry air at 29.92 inches Hg gauge, 70 degrees F.

Scrubber, Fume: A device used to remove contaminants from fume hood exhaust, normally utilizing water.

SDS, Safety Data Sheet: Previously called a MSDS sheet. A safety data sheet is a standardized document that contains occupational safety and health data. The International Hazard Communication Standard (HCS) mandates that chemical manufacturers must communicate a chemical's hazard information to chemical handlers by providing a Safety Data Sheet. SDS's typically contain chemical properties, health and environmental hazards, protective measures, as well as safety precautions for storing, handling, and transporting chemicals.

Self-aware: Is an awareness of self and your relationship to the external world. A self-aware device is not only aware of what it is doing, but is also aware of its surroundings. The programming not only looks at what the device is doing, but also considers what is happening around it.

Sensor Array: Is a group of sensors that are collectively taking multiple readings for a complex view of a situation.

Service Fixture: Item of laboratory plumbing mounted on or fastened to a laboratory fume hood

Setback: Is a setpoint below normal. For example, a system might have a night setback. When it goes into the setback mode, it uses a different set of parameters than it does during day mode.

Shall: When used in a Standard indicates a mandatory feature.

Should: When used in a Standard indicates a recommendation, but is not a mandatory feature.

Slot Velocity: Speed of air moving through fume hood baffle openings.

Smart Hood: A smart hood is one that incorporates a digital controller to manage various functions and performance parameters of the fume hood.

Smart Laboratories: A smart laboratory integrates a number of smart devices that can be controlled with a digital interface.

Smoke Candle: Smoke-producing device used to allow visual observation of airflow.

Smoke Machine: A machine that creates a non-toxic theatrical fog. Used for airflow visualization.

Spot Collector: A small, localized ventilation hood usually connected by a flexible duct to an exhaust fan. This device is not a laboratory fume hood.

Stack: The tube/device on the end of a ventilation system, which disperses exhaust contaminants for dilution by the atmosphere.

Standard Air: Standard Conditions STP Dry air at 70 degrees F, 29.92 in Hg.

Standard Operating Procedure, SOP: This is a formal and written procedure for conducting activities.

Static Pressure: Air pressure in laboratory fume hood or duct, usually expressed in inches of water.

Static Pressure, SP: The pressure developed in a duct by a fan; SP exerts influence in all directions; the force in inches of water measured perpendicular to flow at the wall of the duct; the difference in pressure between atmospheric pressure and the absolute pressure inside a duct, cleaner, or other equipment.

Static Pressure Loss: Measurement of resistance created when air moves through a duct or hood, usually expressed in inches of water.

Suction Pressure: See Static Pressure.

Sulfur Hexafluoride (SF6): Tracer gas widely used for ASHRAE and EN testing. It can be detected with a leak detector to measure loss of containment.

Superstructure: The portion of a laboratory fume hood that is supported by the work surface.

Supplemental Air: Supply or auxiliary air delivered to a laboratory fume hood to reduce room air consumption.

Supply Diffuser: An outlet in the room for the supply air. There are many types, but around fume hoods, a low velocity type is preferred.

Synchronized Supply Fume Hood: This is a new class of fume hood that brings supply air directly into the fume chamber and both the supply and the exhaust are controlled by independent VAV valves so that the proper balance with the fume chamber can be maintained. As the exhaust volume goes up and down, the supply is synchronized to track it.

Synthetic Sensors: General-purpose sensing, wherein a single, highly capable sensor can indirectly monitor a large context, without direct instrumentation of objects. A super sensor that can indirectly monitor a number of activities and provide actionable output.

Table Top Hood: A small, spot ventilation hood for mounting on table tops. Used primarily in educational laboratories. This device is not a laboratory fume hood.

Temperature Gradient: A temperature gradient is a physical quantity that describes in which direction and at what rate the temperature changes the most rapidly around a particular location.

Fume Hood Test Room or Test Lab: This is a space designed and equipped based on criteria in the testing specifications such as ASHRAE110 and EN 14175. This space is designed to have near perfect room conditions so what is being tested is the hood's performance independent of the room.

Thermal Anemometer: A device for measuring fume hood velocity utilizing the principle of thermal cooling of a heated element as the detection element.

Threshold Limit Valve-Time Weighted Average (TLV-TWA): The time-weighted average concentration for a normal 8-hour workday or 40-hour week, to which nearly all workers may be repeatedly exposed, day after day, without adverse effect.

Titanium Tetrachloride: Chemical that generates white fumes used in testing laboratory fume hoods.

Total Pressure, TP: The pressure exerted in a duct as the Algebraic sum of the static pressure and the velocity pressure.

Total Suspended Particulate Matter: The mass of particles suspended in a unit volume of air (typically one cubic meter) when collected by a high-volume sampler.

Transfer air: Air that moves between spaces in a building, driven by the ventilation system.

Transport Velocity: Minimum speed of air required to support and carry particles in an air stream.

Turbulent Flow: Airflow characterized by transverse velocity components, as well as velocity in the primary direction of flow in a duct; mixing velocities.

Turndown: See setback

TWA, Time Weighted Average: The average exposure at the breathing zone.

Ultra-Low Flow Fume Hoods: This is a class of product that is designed to operate at very low average face velocities. For years the benchmark was 100fmp/0.5mps then the low flow hoods were operating at 60fmp/0.3mps. Ultra-low flow goes below that. Often times supplemental air is injected into the fume chamber in an effort to increase containment.

Variable Air Volume - two-position ventilation system: A constant air volume ventilation system (sometimes also referred to as a "two-position variable air volume system") that is designed to provide two separate levels of airflow. The higher level of airflow is provided when a facility is normally occupied such as during regular work hours. The lower level of airflow is utilized during unoccupied times (e.g., nighttime, holidays, etc.) when ventilation needs and internal loads require less airflow.

Variable Air Volume (VAV): In a HVAC system, the supply air volume is varied by dampers or fan speed controls to maintain the temperature; in hoods, the exhaust air is varied to reduce the amount of air exhausted. If the fume hood is going to operate in VAV mode, then the room supply has to also be VAV to keep the room in balance.

Variable air volume (VAV) ventilation system: A type of HVAC system specifically designed to vary the amount of conditioned air supplied and exhausted from the spaces served. The amount of air supplied and intended to meet (but not exceed) the actual need of a space at any point in time. In general, the amount of air that is needed by a space is determined by the required rate and the amount of airflow necessary to maintain comfortable conditions (temperature and humidity). In laboratories, we also consider the required number of ACH to minimize the risk of exposure to the users.

Variable Frequency Drive (VFD): Is a type of adjustable-speed drive used in electro-mechanical drive systems to control AC motor speed by varying motor input frequency. These are very common on fans used for exhaust and supply air.

Variable Volume Hood: A hood designed so the exhaust volume is varied in proportion to the opening of the hood face by changing the speed of the exhaust blower or by operating a damper or control valve in the exhaust duct.

Velocity, V: Magnitude of air motion perpendicular to the plane of the airflow cross section.

Velocity Pressure: Pressure caused by moving air in a laboratory fume hood or duct, usually expressed in inches of water.

Vena Contracta: is the point in a fluid stream where the diameter of the stream is the least, and fluid velocity is at its maximum, such as in the case of a stream issuing out of a nozzle (orifice). It is a place where the cross-section area is minimum.

Ventilated Enclosure: These are often fume hood like devices that are used for exhausting laboratory air but do not meet the definition or performance criteria of a laboratory fume hood.

Visualization of Airflow: Since airflow can't be viewed directly, we have developed various methods to help visualize how the airflow is behaving. The most common is theatrical smoke or fog, also smoke bombs, smoke candles can be used. Using a modified laser beam can further enhance the visualization by projecting it onto the smoke. This is helpful when trying to understand air patterns within the fume hood.

Volume Flow Rate, Q: The quantity of air flowing in cubic feet per minute, cfm, scfm, acfm.

Volumetric airflow rate: The rate of airflow expressed in terms of volume (cubic feet or liters) per unit of time. These are commonly expressed as cubic feet per minute (cfm) in USCS units or liters per second (l/s) in SI units. (Also see room ventilation.)

Vortex: A mass of whirling air, a rotating air mass that often forms insides fume hoods. The most common is a horizontal vortex that forms at the top of the fume chamber behind the sash. But in combination sashed hoods, it is also common to see vertical vortex that form behind the closed section of sash. The performance issue with vortex is that they can collect contaminants and spin them near the sash opening increasing the chance for loss of containment.

Walk-in hood: See floor-mounted hood.

Weather Cap: Device used at the top of an exhaust stack to prevent rain from entering the stack end. Because of the need for a high velocity discharge, only specially designed rain guards should be used. Rain caps that cover the stack are not acceptable.

Work Space: The part of the fume hood interior where apparatus is set up and fumes are generated. It is normally confined to a space extending from six inches (15.2 cm) behind the plane of the sash(es) to the face of the baffle, and extending from the work surface to a plane parallel with the top edge of the access opening.

Work Surface: The surface that a laboratory fume hood is located on and supported by a base cabinet. In the fume chamber, the surface is recessed to contain spills.

Made in United States
Orlando, FL
04 November 2026

90202237R00085